TCP 2014

Michiharu Wada • Peter Schury
Yuichi Ichikawa
Editors

TCP 2014

Proceedings of the 6th International Conference
on Trapped Charged Particles and Fundamental
Physics, held in Takamatsu, Japan, 1-5,
December 2014

 Springer

Editors
Michiharu Wada
RIKEN Nishina Center
  for Accelerator-Based Science
Saitama, Japan

Peter Schury
RIKEN Nishina Center
  for Accelerator-Based Science
Saitama, Japan

Yuichi Ichikawa
RIKEN Nishina Center
  for Accelerator-Based Science
Saitama, Japan

ISBN 978-3-319-87115-8   ISBN 978-3-319-61588-2 (eBook)
DOI 10.1007/978-3-319-61588-2

Printed on acid-free paper

This Springer imprint is published by Springer Nature
The registered company is Springer International Publishing AG
The registered company address is: Gewerbestrasse 11, 6330 Cham, Switzerland

With contributions from

Safdar Ali • Zoran Andelkovic • S. Ando • T. Aoki • H. Arikawa • K. Asahi
M.D. Ashkezari • Samuel Ayet San Andres • G. Ban • Steffi Bandelow
M. Baquero-Ruiz • E. Bennett • W. Bertsche • C.P. Bidinosti • Gerhard Birkl • R. Burch
C. Burrows • E. Butler • A. Capra • C.L. Cesar • M. Charlton • M. Chikamori
C. Couratin • P. Delahaye • Timo Dickel • M. Diermaier • Jens Dilling • R. Dunlop
D. Durand • V.A. Dzuba • M.S. Ebrahimi • S. Eriksson • N. Evetts • X. Fabian • B. Fabre
J. Fajans • S. Federmann • Svetlana Fedotova • B. Fenker • P. Finlay • V.V. Flambaum
X. Fléchard • T. Friesen • T. Fujita • Makoto C. Fujiwara • K. Fuke • K. Fukushima
T. Fukuyama • C. Funayama • T. Furukawa • Hans Geissel • D.R. Gill • Florian Greiner
A. Gutierrez • J.S. Hangst • Volker Hannen • Hideaki Hara • Hirohisa Hara • K. Harada
W.N. Hardy • Shuichi Hasegawa • A. Hatakeyama • T. Hayamizu • M.E. Hayden
Frank Herfurth • H. Higaki • E. Hikota • C. Hirao • Masanari Ichikawa • Y. Ichikawa
K. Imamura • T. Ino • T. Inoue • C.A. Isaac • T. Ishikawa • K. Ito • M. Itoh
Christian Jesch • S. Jonsell • Kyunghun Jung • Y. Kanai • Daiji Kato • K. Kato
Hirokazu Kawamura • T. Kobayashi • L. Köhler • S. Kojima • Kristian König
Nikita Kotovskiy • Susumu Kuma • L. Kurchaninov • N. Kuroda • Johannes Lang
Kyle G. Leach • M. Leali • E. Liénard • Wayne Lippert • A. Little • Bernhard Maaß
Kirk W. Madison • N. Madsen • C. Malbrunot • J. Mario Michan • A. Martin
Franklin Martinez • Gerrit Marx • V. Mascagna • O. Massiczek • Takahiko Masuda
J. Mathis • Y. Matsuda • Y. Matsuo • F. Mauger • J.T.K. McKenna • M. Mehlman
D. Melconian • S. Menary • A. Méry • K. Michishio • Yuki Miyamoto • T. Mizutani
A. Mohri • Takamasa Momose • K. Moriya • Izumi Murakami • Tobias Murböck
D.J. Murtagh • H. Nagahama • Y. Nagashima • Y. Nagata • Nobuyuki Nakamura
Itsuo Nakano • S.C. Napoli • O. Naviliat-Cuncic • Dennis Neidherr • P. Nolan
Wilfried Nörtershäuser • Chiaki Ohae • Y. Ohshima • Y. Ohtomo • M. Ohtsuka
Kunihiro Okada • H. Okamoto • T. Okano • K. Olchanski • A. Olin • G.G. Paulus
Wolfgang R. Plaß • Gene Polovy • B. Pons • T. Porobic • P. Pusa • G. Quéméner
W. Quint • B. Radics • C.Ø. Rasmussen • S. Ringleb • E. Lodi Rizzini • F. Robicheaux
R.L. Sacramento • K. Sakamoto • Y. Sakamoto • Hiroyuki A. Sakaue • Y. Sakemi
S. Sakurai • E. Sarid • Noboru Sasao • T. Sato • C. Sauerzopf • Christoph Scheidenberger
Stefan Schmidt • Lutz Schweikhard • N. Severijns • P.D. Shidling • Erina Shimizu
T. Shimoda • Devin Short • D.M. Silveira • C. So • Th. Stöhlker • S. Stracka • K. Suzuki
T. Suzuki • M. Tajima • Minoru Tanaka • J. Tarlton • T.D. Tharp • J.C. Thomas
R.I. Thompson • M. Tona • P. Tooley • H.A. Torii • M. Tsuchiya • M. Turner
A. Uchiyama • H. Ueno • Satoshi Uetake • S. Ulmer • D.P. van der Werf • S. Van Gorp
Albert Vass • Ph. Velten • L. Venturelli • Manuel Vogel • Jonas Vollbrecht
D. von Lindenfels • Gleb Vorobjev • Michiharu Wada • Tetsuya Watanabe
Christian Weinheimer • E. Widmann • M. Wiesel • Hermann Wollnik • B. Wünschek
J.S. Wurtele • Yuta Yamamoto • Norimasa Yamamoto • Y. Yamazaki • X.F. Yang
Mikhail I. Yavor • Akihiro Yoshimi • Koji Yoshimura • Motohiko Yoshimura
A.I. Zhmoginov • J. Zmeskal • N. Zurlo

Contents

Hyperfine Interact (2015) 235:7–11
DOI 10.1007/s10751-015-1208-y

Preface

Michiharu Wada[1,2]

Published online: 12 January 2016

The 6th International Conference on Trapped Charged Particles and Fundamental Physics (TCP2014) took place at Kagawa International Hall in Sunport Takamatsu Symbol Tower of Takamatsu city on the Japanese island of Shikoku during December 1–5, 2014, and was hosted by the RIKEN Nishina Center for Accelerator-based Science. The conference was the latest in a series of very successful meetings beginning in Lysekil (Sweden) in 1994 and continued in Asilomar (USA) in 1998, Wildbad Kürth (Germany) in 2002, Parksville (Canada) in 2006, and Saariselkä (Finland) in 2010. The 2014 meeting followed the theme of previous events by being close to the ocean and being isolated in a remote place to keep the participants together.

We would remind you that right after the last meeting in Finland, many participants were stranded in Helsinki for a week due to the Icelandic volcano explosion. No aircrafts were operated within the entirety of Europe. This disaster made us aware of the power of nature. Following this incident, the Great East-Japan Earthquake struck on March 11, 2011. The quakes and following Tsunami, and the accident in the nuclear power plants, hurt citizens in Japan and changed people's minds significantly. We scientists were also concerned as to whether we could continue fundamental research that required big budgets and large power consumption. Nuclear physicists were even criticized, although nuclear engineering and nuclear physics are totally different fields. If we take a deep look at history, however, nuclear fuel was synthesized by the r-process in the universe, which is one of the major subjects of our physics today, and atomic energy is the most important social application born from nuclear physics. We nuclear physicists, however, cannot escape from original sin for opening Pandora's box. We therefore worked as best we could as radiation workers with expertise in nuclear physics to help deal with the aftermath of the disaster. Although our

Proceedings of the 6th International Conference on Trapped Charged Particles and Fundamental Physics (TCP 2014), Takamatsu, Japan, 1–5 December 2014

✉ Michiharu Wada
mw@riken.jp

1   RIKEN Nishina Center for Accelerator-based Science, Wako, Japan

2   Present address: Wako Nuclear Science Center, IPNS, KEK, Wako, Japan

contributions were very limited in such a nuclear crisis, we hoped for swift recovery from the tragedy. Thanks to the great efforts of a variety of people—those involved in management of the plants, decontamination of wide areas within Tohoku, control of agricultural products, medical service for citizen, and so on—we could avoid the worst scenario of the disaster. It allowed us to continue research activities and, thus, to organize the TCP2014 conference in late 2014.

Prior to the conference, a school (TCP school) for young researchers was organized during November 28–29, 2014, at the Nishina Hall of RIKEN, where inspiring lectures were given by S. Ulmer, R. Thompson, R. Hayano, H. Katori, V. Dzuba, Yu. Litvinov, and W. Nörtershäuser. About 60 participants enrolled in the school, half of whom went also to Takamatsu.

The conference was structured into nine sessions, covering the topics of Anti-Hydrogen, Ion Traps for HCI, Storage Rings, Application of Particle Trapping, Fundamental Interactions and Symmetries, Quantum and QED Effects, and Precision Spectroscopy and Frequency Standards. Each topic consisted of a keynote talk, invited talks and contributed talks. Many new developments and exciting results were presented in these sessions.

The conference started on Monday morning with a warm welcome by H. En'yo, the director of RIKEN Nishina Center. The first scientific session was initiated with a keynote talk by Y. Yamazaki, providing a review of anti-matter science. Since the last conference, the topic of Anti-Hydrogen advanced the most; anti-hydrogen was routinely synthesized and the competition is now focused on the precision spectroscopy of anti-matter. The Storage Ring sessions had two keynote talks, with Yu. Litvinov discussing large magnetic rings and H. Schmidt discussing small electrostatic rings. H. Wollnik reviewed the history of mass spectrometers in the Application of Particle Trapping session. In this session, many advanced features of Penning trap mass spectrometers were presented and newly available multi-reflection time-of-flight mass spectrographs at three different laboratories (RIKEN, CERN, and GSI) were also introduced. Such spectrographs were also under preparation at TRIUMF, MSU, ANL and IBS. Various studies on highly charged ions were discussed in the Ion Traps for HCI session. A highlight of this session was a sympathetically cooled HCI crystal in a linear Paul trap presented by O. Versalato. It could become a next-generation frequency standard as well as a platform for research in physics beyond the standard model. D. Leibfried gave a lecture on basics of quantum information processing in the Quantum and QED Effect session. Since NNP2014 (International Workshop on Non Neutral Plasma 2014) was held at the same time and location, a joint session was held on Wednesday morning, where E. Cornell gave the keynote talk on electron EDM studies with molecular ions. Several other talks on the topics of Anti-Hydrogen and Precision Spectroscopy and Frequency Standard were presented as well. Topics on Fundamental Interactions and Symmetries had several invited talks; some of them even used neutral particles. In total, 45 invited talks and 23 oral presentations were presented. In addition, a well-attended poster session was held on Monday evening. Among 29 posters, poster awards were given to four young participants: A. Gurierrez, J. Tarlton, F. Arai, and C. Funayama.

Overall, the conference was quite successful. Steady advances from the previous conference were seen for all topics. It is suitable to declare that the TCP conference series remains the flagship conference of these fields. During the conference, an International Advisory Committee (IAC) meeting was held to discuss the next TCP conference, which will be held in 2018. It was decided that it will take place in USA and be hosted by Michigan State

University. At this point we would like to thank the members of the IAC for their help in putting together the program of the conference.

It must be noted that the volume of these proceedings is small, with only 23 papers, while 94 contributions were presented in the conference. It is unfortunate, however, that the motivation for writing such proceedings is diminishing. Some organizations even prohibit their employees to submit proceedings due to possible overlaps with parts of their original papers. This kind of infringement becomes more serious today. On this basis, we think it is time to seriously discuss whether or not proceedings are needed.

Acknowledgments We would like to thank the main sponsors, Kagawa Prefecture, Takamatsu Convention & Visitors Bureau (TCVB) and RIKEN Nishina Center for Accelerator-Based Science for organizing the conference as well as for financial support. In particular, the contributions of Y. Sato of TCVB were indispensable for coordination in Takamatsu. Additional financial support from CANBERRA Industries Inc., FUJIKIN Corp., NEC Tokin Corp., OPTIMA Co. Ltd., p-ban.com Corp., REPIC Corp., Tokyo Electronics Co. Ltd., Spectra-Physics Co. Ltd., and Thorlabs Japan Inc. were also appreciated. The local conference team (M. Wada, P. Schury, Y. Ichikawa, Y. Ito, and S. Naimi) was very well supported by the secretaries, N. Kiyama and E. Isogai.

Hyperfine Interact (2015) 235:1–5
DOI 10.1007/s10751-015-1207-z

Chairs, Committees, Sponsors and Participants

Published online: 18 January 2016

International Advisory Committee

Juha Äystö	(Helsinki Institute of Phyics)
Georg Bollen	(NSCL/Michigan State Univ.)
John Bollinger	(NIST)
Klaus Blaum	(MPIK)
Andrzej Czrnecki	(University of Alberta)
Jens Dilling	(TRIUMF)
Michael Doser	(CERN)
Gerald Gabrielse	(Harvard University)
Ari Jokinen	(University of Jyväskylä)
Klaus Jungmann	(KVI)
H-Jürgen Kluge	(GSI)
Youngkyun Kim	(IBS)
Xinwen Ma	(IMP)
Oscar Naviliat-Cuncic	(NSCL/Michigan State Univ.)
Christoph Scheidenberger	(GSI)
Guy Savard	(ANL/University of Chicago)
Nathal Severijns	(University of Leuven)
Reinhold Schuch	(Stockholm University)
David Wineland	(NIST)
Eberhard Widmann	(SMI)
Michiharu Wada	(RIKEN) Chair

Proceedings of the 6th International Conference on Trapped Charged Particles and Fundamental Physics
(TCP 2014), Takamatsu, Japan, 1–5 December 2014

Supporting Organizations

Kagawa Prefecture
Takamatsu Convention and Visitor Bureau
RIKEN Nishina Center for Accelerator-
Based Science

Local Organizing Committee

Toshiyuki Azuma	(RIKEN)
Hiroyuki Higaki	(Hiroshima University)
Yuichi Ichikawa	(RIKEN) Scientific Secretary
Yuta Ito	(RIKEN)
Nobuyuki Nakamura	(University of Elec.-Commun.)
Akira Ozawa	(University of Tsukuba)
Yasuhiro Sakemi	(CYRIC, Tohoku University)
Peter Schury	(RIKEN) Scientific Secretary
Kazuhiko Sugiyama	(Kyoto University)
Hideki Ueno	(RIKEN)
Tomohiro Uesaka	(RIKEN)
Michiharu Wada	(RIKEN) Chair

Secretaries

Noriko Kiyama	(RIKEN)
Emiko Isogai	(RIKEN)

Corporate Sponsors

CANBERRA Industris Inc
FUJIKIN Corp.
NEC TOKIN Corp.
OPTIMA Co., Ltd.
p-ban.com Corp
REPIC Corp.
Tokyo Electronics Co., Ltd.
Spectra-Physics Co., Ltd.
Thorlabs Japan Inc.

Participants

Safdar ALI	(UEC, Japan)
Zoran ANDELKOVIC	(GSI, Germany)
Fumiya ARAI	(University of Tsukuba, Japan)
Toshiyuki AZUMA	(RIKEN, Japan)
Gilles BAN	(LPC Caen, France)
Thomas BAUMANN	(NSCL, USA)
Nikolay BELOV	(Max-Planck-Institut für Kernphysik, Germany)

Michael BLOCK   (GSI, Germany)
Georg BOLLEN   (FRIB, USA)
John BOLLINGER   (NIST, USA)
Thomas BRUNNER   (Stanford University, USA)
Giovanni CERCHIARI   (Max-Planck-Institut für Kernphysik, Germany)

Ankur CHAUDHURI   (Institute for Basic Science, Korea)
Xiangcheng CHEN   (GSI, Germany)
Jason CLARK   (Argonne National Laboratory, USA)
Danial COMPARAT   (CNRS, France)
Eric CORNELL   (NIST and JILA, USA)
Pierre DERAHAYE   (GANIL, FRANCE)
Timo DICKEL   (GSI, Germany)
Michael DOSER   (CERN, Switzerland)
Pierre DUPRE   (CSNSM-IN2P3, France)
Vladimir DZUBA   (University of New South Wales, Australia)
Martin EIBACH   (NSCL, USA)
Sergey ELISEEV   (Max-Planck-Institut für Kernphysik, Germany)

Hideto EN'YO   (RIKEN, Japan)
Tommi ERONEN   (University of Jyväskylä, Finland)
Xavier FABIAN   (LPC Caen, France)
Hiroto FUJISAKI   (Kyoto University, Japan)
Tomomi FUJITA   (Osaka University, Japan)
Makoto FUJIWARA   (TRIUMF, Canada)
Kiyokazu FUKE   (Institute for Molecular Science, Japan)
Chikako FUNAYAMA   (Tokyo Institute of Technology, Japan)
Takeshi FURUKAWA   (Tokyo Metropolitan University, Japan)
Dmitry GLAZOV   (St. Petersburg State University, Russia)
Andrea GUTIERREZ   (University of British Columbia, Canada)
James HARRIES   (Japan Atomic Energy Agency, Japan)
Shuichi HASEGAWA   (University of Tokyo, Japan)
Frank Herfurth   (GSI, Germany)
Masaki HORI   (Max-Planck-Institut für Kernphysik, Germany)

Pavel HRMO   (Imperial College London, United Kingdom)
Yuichi ICHIKAWA   (RIKEN, Japan)
Kang-Bin IM   (RISP, Korea)
Takeshi INOUE   (Tohoku University, Japan)
Takahisa ITAHASHI   (Osaka University, Japan)
Kiyokazu ITO   (Hiroshima University, Japan)
Yuta ITO   (RIKEN, Japan)
Munendra JAIN   (BRCMCET BAHAL, India)
Ari JOKINEN   (University of Jyväskylä, Finland)
Elena JORDAN   (Max-Planck-Institut für Kernphysik, Germany)

Hirokazu KAWAMURA   (Tohoku University, Japan)
Taehyun KIM   (SK Telecom, Korea)
H.-Jürgen KLUGE   (GSI, Germany)

Naofumi KURODA	(University of Tokyo, Japan)
Ania KWIATKOWSKI	(TRIUMF, Canada)
Dietrich LEIBFRIED	(NIST, USA)
Etienne LIENARD	(LPC Caen, France)
Yuri LITVINOV	(GSI, Germany)
Takahiko MASUDA	(Okayama University, Japan)
Michael MEHLMAN	(Texas A&M University, USA)
Dan MELCONIAN	(Texas A&M University, USA)
Szilard NAGY	(Max-Planck-Institut für Kernphysik, Germany)
Sarah NAIMI	(RIKEN, Japan)
Nobuyuki NAKAMURA	(University of Electro-Communications, Japan)
Wilfried NOERTERHAESER	(TU Darmstadt, Germany)
Naoki NUMADATE	(Sophia University, Japan)
Kunihiro OKADA	(Sophia University, Japan)
Natalia ORESHKINA	(Max-Planck-Institut für Kernphysik, Germany)
Young-Ho PARK	(RISP, Korea)
Patrice PEREZ	(CEA/Saclay, France)
Wolfgang PLASS	(GSI, Germany)
Petteri PUSA	(University of Liverpool, United Kingdom)
Matthew REDSHAW	(Central Michigan University, USA)
Mikael REPONEN	(RIKEN, Japan)
Ryan RINGLE	(NSCL, USA)
Christian ROOS	(IQOQI Innsbruck, Austria)
Marco ROSENBUSCH	(Universität Greifswald, Germany)
Naohito SAITO	(KEK, Japan)
Hiroyuki SAKAUE	(National Institute for Fusion Science, Japan)
Yasuhiro SAKEMI	(Tohoku University, Japan)
Tomoya SATO	(Tokyo Insitute of Technology, Japan)
Henning SCHMIDT	(Stockholm University, Sweden)
Reinhold SCHUCH	(Stockholm University, Sweden)
Peter SCHURY	(RIKEN, Japan)
Stefan SCHWARZ	(NSCL, USA)
Lutz SCHWEIKHARD	(University of Greifswald, Germany)
Vladimir SHABAEV	(St. Petersburg State University, Russia)
Martin SIMON	(Stefan Meyer Institute, Austria)
Prithvi SINGH	(Sir Padampat Singhania University, India)
Christian SMORRA	(CERN, Switzerland)
Matthew STERNBERG	(University of Washington, USA)
Cody STORRY	(York University, United Kingdom)
Sven STURM	(Max-Planck-Institut für Kernphysik, Germany)
Kazuhiko SUGIYAMA	(Kyoto University, Japan)
Yusuke TAKADA	(Sophia University, Japan)
Keiji TAKAHISA	(RCNP, Osaka University, Japan)
Aiko TAKAMINE	(Aoyama Gakuin University, Japan)

James TARLTON	(University of Oxford, United Kingdom)
Kenji TOYODA	(Osaka University, Japan)
Tomohiro UESAKA	(RIKEN, Japan)
Stefan ULMER	(RIKEN, Japan)
Oscar VERSOLATO	(Advanced Research Center for Nanolithography, Netherlands)
Manuel VOGEL	(GSI, Germany)
Michiharu WADA	(RIKEN, Japan)
Masanori WAKASUGI	(RIKEN, Japan)
Lorenz WILLMAN	(University of Groningen, Netherlands)
Robert WOLF	(Max-Planck-Institut für Kernphysik, Germany)
Hermann WOLLNIK	(New Mexico State University, USA)
Yoshitaka YAMAGUCHI	(RIKEN, Japan)
Ke YAO	(IMP, Fudan University, China)
Yu Hu ZHANG	(IMP, Lanzhou, China)
Chuan ZHENG	(IMP, Fudan University, China)
Ge ZHUANG	(RIKEN/IMP, Japan)

Hyperfine Interact (2015) 236:1–7
DOI 10.1007/s10751-015-1198-9

Precision measurements with LPCTrap at GANIL

E. Liénard[1] · G. Ban[1] · C. Couratin[2] · P. Delahaye[3] · D. Durand[1] · X. Fabian[1] ·
B. Fabre[4] · X. Fléchard[1] · P. Finlay[2] · F. Mauger[1] · A. Méry[5] ·
O. Naviliat-Cuncic[6] · B. Pons[4] · T. Porobic[2] · G. Quéméner[1] · N. Severijns[2] ·
J. C. Thomas[3] · Ph. Velten[2]

Published online: 14 July 2015
© Springer International Publishing Switzerland 2015

Abstract The experimental achievements and the results obtained so far with the LPCTrap
device installed at GANIL are presented. The apparatus is dedicated to the study of the weak
interaction at low energy by means of precise measurements of the $\beta - \nu$ angular correlation
parameter in nuclear β decays. So far, the data collected with three isotopes have enabled
to determine, for the first time, the charge state distributions of the recoiling ions, induced
by shakeoff process. The analysis is presently refined to deduce the correlation parameters,
with the potential of improving both the constraint deduced at low energy on exotic tensor
currents ($^6\mathrm{He}^{1+}$) and the precision on the V_{ud} element of the quark-mixing matrix ($^{35}\mathrm{Ar}^{1+}$
and $^{19}\mathrm{Ne}^{1+}$) deduced from the mirror transitions dataset.

Keywords Ion trapping · Correlation in nuclear β–decay · Test of weak interaction ·
Shakeoff process

Proceedings of the 6th International Conference on Trapped Charged Particles and Fundamental
Physics (TCP 2014), Takamatsu, Japan, 1-5 December 2014

✉ E. Liénard
  lienard@lpccaen.in2p3.fr

[1] LPC CAEN, ENSICAEN, Université de Caen, CNRS/IN2P3, Caen, France

[2] Instituut voor Kern- en Stralingsfysica, KU Leuven, 3001 Leuven, Belgium

[3] GANIL, CEA/DSM-CNRS/IN2P3, Caen, France

[4] CELIA, Université Bordeaux, CNRS, CEA, Talence, France

[5] CIMAP, CEA/CNRS/ENSICAEN, Université de Caen, Caen, France

[6] NSCL and Department of Physics and Astronomy, MSU, East-Lansing, MI, USA

1 Introduction

Correlation measurements in nuclear β decays enable to probe the structure of the weak interaction, complementarily to high energy physics experiments [1, 2]. In particular, the study of the angular correlation between the two leptons gives access to the parameter $a_{\beta\nu}$ sensitive to the existence of exotic currents, scalar or tensor, beyond the $V - A$ structure of the Standard Model (SM). This parameter depends quadratically on the coupling constants associated to the different currents considered in the weak interaction. In the frame of the SM, for allowed nuclear decays it is given by:

$$a_{\beta\nu} = \frac{1 - \rho^2/3}{1 + \rho^2}$$

where $\rho = \frac{C_A M_{GT}}{C_V M_F}$ is the mixing ratio of the transition, C_A and C_V are the coupling constants associated to the axial-vector and vector currents respectively, M_F and M_{GT} are the Fermi (F) and Gamow-Teller (GT) nuclear matrix elements. As a consequence, $a_{\beta\nu}$ equals $1 (-1/3)$ for a pure F (GT) transition. Any deviation from these values would imply either a departure from the allowed approximation or the presence of new physics beyond the SM. However, the distribution of events also depends on the Fierz term, b, which arises from the interference between exotic and standard currents and is therefore null in the SM. This particularity enables to consider that the effective parameter that is determined in a $\beta - \nu$ angular correlation experiment is actually [3]:

$$\tilde{a}_{\beta\nu} = a_{\beta\nu}/\left(1 + \left\langle b\frac{m}{E_e}\right\rangle\right)$$

where the brackets $<>$ mean a weighted average over the measured part of the β spectrum, m and E_e are respectively the mass and the total energy of the β particle.

The first candidate studied with LPCTrap at GANIL was ^{6}He [4], which is an ideal case to probe the tensor components. Until now only one experiment, performed in 1963, has reached a relative precision at the level of 1 % (1σ) in this decay, yielding $a_{\beta\nu} = -0.3308(30)$ [5, 6]. This result contributed significantly to fix the current limits on tensor contributions deduced from the correlation measurements performed at low energy [3].

For mirror transitions, the measurement of $a_{\beta\nu}$ also allows for a precise determination of the mixing ratio ρ. This parameter, combined with precise half-life, branching ratio and masses, can be used to compute the V_{ud} element of the Cabibbo-Kobayashi-Maskawa quark-mixing matrix [7]:

$$V_{ud}^2 = \frac{4.794 \times 10^{-5}}{(Ft_{1/2})G_F^2 |M_F|^2 (1 + \Delta_R)\left(1 + \frac{f_A}{f_V}\rho^2\right)} \tag{1}$$

where $Ft_{1/2}$ is the corrected ft-value of the transition, G_F is the Fermi coupling constant, Δ_R is a transition-independent radiative correction and f_A (f_V) is the statistical rate function computed for the GT (F) component. In the T = 1/2 mirror transitions [8], the mixing ratio is always the least known parameter [9]. In the case of ^{35}Ar and ^{19}Ne, all parameters involved in the determination of V_{ud}, except ρ, are presently known with relative precisions below the 10^{-4} level. Accurate correlation measurements in their decays would then enable to significantly improve the current value of V_{ud} deduced from the mirror transitions database [7]: $V_{ud} = 0.9719(17)$, which is a factor of 10 less precise than the result deduced from the pure F transitions [10]. ^{35}Ar and ^{19}Ne are two species also studied with LPCTrap [11].

Table 1 Performances of LPCTrap during the last experiments with ^{6}He, ^{35}Ar and ^{19}Ne (see text for details)

Beam (year of exp.)	I_{beam} (pps)	Buffer gas	$\varepsilon_{LPCTrap}$ cycle $= 200$ ms	Trapped radio-active ions/cycle	N_{coinc}
^{6}He^{1+} (2010)	1.5×10^8	H_2	5×10^{-4}	1.5×10^4	1.2×10^6
^{35}Ar^{1+} (2012)	3.5×10^7	He	4×10^{-3}	2.5×10^4	1.5×10^6
^{19}Ne^{1+} (2013)	2.5×10^8	He	9×10^{-4}	4.5×10^4	1.3×10^5

I_{beam} is the beam intensity at the entrance of LPCTrap; the buffer gas is used in the RFQ and the measurement trap to cool the ions; $\varepsilon_{LPCTrap}$ is the transmission efficiency of LPCTrap

In all experiments performed in the past, $a_{\beta\nu}$ was always deduced from the distribution of a kinematic parameter of the recoiling daughter ion (RI), since the ν detection is not efficient. Because of the very low kinetic energy of the RI, traps offer an ideal environment to confine the radioactive source and to ensure minimal disturbance for the RI motion [12–16]. The central element of LPCTrap is a transparent Paul trap [17], allowing the detection in coincidence of the β particle and the RI. The setup is installed at the low energy beam line, LIRAT, of the GANIL/SPIRAL facility.

2 Performances of LPCTrap

At GANIL, the low energy radioactive beams delivered to LPCTrap are provided by the SPIRAL ECR source with a typical energy dispersion of 20 eV at 10 keV kinetic energy. A radio-frequency quadrupole cooler buncher (RFQCB) is then used to reduce the beam emittance and to produce ion bunches. The RFQCB is connected to the transparent Paul trap by a short line with dedicated beam optics and diagnostics. A telescope made of a double-sided silicon strip detector and a thick plastic scintillator is used to detect the β particles while the RI's are detected in coincidence thanks to a micro-channel plates position sensitive system. The detectors are set in a back-to-back configuration, combining the highest statistics and the better sensitivity to a tensor component. The LPCTrap setup is described in detail in [11, 14] and references therein. Some important features are summarized here:

- The trigger of an event is given by the detection of a particle in the β telescope. Then many parameters are measured, such as the RI time-of-flight (ToF), the positions of the two particles, the trap RF and the timestamp of the decay in the measurement cycle. The set of distributions enables to reduce the background and to fix and control systematic parameters of the experiment during the off-line analysis.
- The current version of the setup contains a recoil ToF spectrometer which allows the separation of the different charge states of the RI, due to the shakeoff process. This spectrometer makes LPCTrap a unique setup to measure the charge state distribution of the RI after the β decay of singly charged ions.
- Until now, a measurement cycle of 200 ms (100 ms in the first experiment [14]) was used: an ion bunch is extracted and sent to the Paul trap each 200 ms. Actually the ions are kept in the trap during 160 ms and then extracted to measure the background during the remaining 40 ms, which is thus controlled continuously during the experiment.

The performances of LPCTrap are summarized in Table 1, for the last experiments performed with ^{6}He, ^{35}Ar and ^{19}Ne. The last column (N_{coinc}) gives the number of "good"

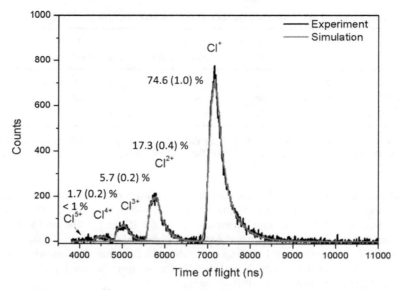

Fig. 1 ToF distribution of the ^{35}Cl RI resulting from ^{35}Ar^{1+} decay

coincidences accumulated in some days, which includes a complete detection of the two particles (energy and position) when the ion cloud in the trap is at equilibrium (a buffer gas is injected in the trap to cool the ions), and the subtraction of the remaining background. This number also depends on many parameters such as the geometrical detection efficiency (4.5×10^{-4}), the emission anisotropy of the β's and the RI in the decay, the rate of recoil neutrals. The lower statistics for ^{19}Ne is due to the long half-life and several technical problems on the beam production encountered during the experiment.

3 Results

3.1 The charge state distributions

The RI spectrometer of LPCTrap has enabled to determine for the first time the experimental value of the shakeoff probability in the decay of ^{6}He^{1+} ions [18]. This system, with one single electron, is an ideal case to test the sudden approximation commonly used in the theoretical descriptions of shakeoff processes. Our measurements have validated this fundamental approximation, yielding an experimental shakeoff probability, $P_{SO} = 0.02339(36)$ given at 1σ, in excellent agreement with its theoretical prediction, $P_{th} = 0.02322$.

In the case of ^{35}Ar^{1+}, several electrons are involved, and such a system has revealed the importance of other processes such as the Auger emission [19]. The experimental charge state distribution of the resulting ^{35}Cl ions (Fig. 1) can indeed be reproduced by calculations only if single and multiple Auger decays, subsequent to inner-shell shakeoff, are explicitly taken into account. For ^{19}Ne^{1+}, preliminary calculations predict a significantly lower effect due to Auger emission process, leading to lower yields for the highest charge states. Such a behavior is qualitatively observed in the ToF distribution of the ^{19}F ions (Fig. 2), which is more strongly dominated by the two first charge states than the ^{35}Cl ionic counterparts in

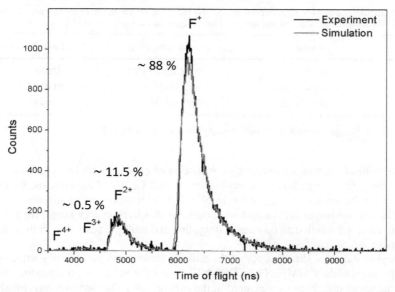

Fig. 2 ToF distribution of the ^{19}F RI resulting from ^{19}Ne^{1+} decay

the ^{35}Ar^{1+} decay. The detailed analysis of these data is currently in progress to determine accurately the experimental charge state distribution, including the neutrals, and to achieve a more constraining comparison with upgraded calculations including recoil effects and possible shakeup contributions.

3.2 The $\beta - \nu$ angular correlation parameters

The extraction of $a_{\beta\nu}$ from the data requires realistic simulations of the experiments, containing a statistics significantly larger than the number of coincidences collected. The parameter is deduced from a fit of the RI ToF distribution using a linear combination of two distributions simulated with different values of $a_{\beta\nu}$ [14]. At least three parameters are left free in the fit: the value of $a_{\beta\nu}$, the total number of events and the distance, d, between the RI detector and the center of the Paul trap. Indeed, the last parameter is by far determined more precisely using the data themselves than a specific geometrical measurement. The two parameters, $a_{\beta\nu}$ and d, are correlated such that the minimization is completely unambiguous (see Fig. 13 in [14]).

For ^{6}He, a first value at a relative precision of 3 % (1σ) has been deduced from the data collected during a first experiment with limited statistics [14]:

$$a_{\beta\nu} = -0.3335(73)_{\text{stat}}(75)_{\text{syst}} \qquad (2)$$

The systematic error is largely dominated by the spatial distribution of the ion cloud inside the trap (91 %), related to its temperature, and determined in an off-line experiment, using a ^{6}Li source [20]. Simulations have shown that this parameter could also be determined with higher precision, if considered as a free parameter of the fit of the RI ToF distribution [21]. Indeed the shape of the leading edge of this distribution depends strongly on the cloud temperature and not on the value of $a_{\beta\nu}$. Finally all the dominant contributions to the final systematic error $(\Delta a)_{\text{syst}}$, except the β scattering in the device, could be determined by

5

Table 2 Statistical precision expected on $a_{\beta\nu}$ in the three nuclei studied at GANIL with LPCTrap (last column). SM and current values of $a_{\beta\nu}$ are given for comparison (see text for details)

Isotope	SM value	Current value [Ref]	Projected 1σ precision
^{6}He	−0.3333	−0.3308(30) [5, 6]	0.0015
^{35}Ar	0.9004[a](16)	0.97(14) [23]	0.0013
^{19}Ne	0.0438[b](8)	0.00(8) [23]	0.0046

[a]From Ref. [8]; [b]adapted from Ref. [8] with the new world average of $T_{1/2}$ [24]

the data themselves, with consequently a reduction of $(\Delta a)_{\text{syst}}$ with higher statistics. Concerning the β scattering, the effect was estimated with GEANT4 simulations, leading to a relative contribution to $(\Delta a)_{\text{syst}}$ at the level of 0.6 %, which remains still reasonable. To reduce it, new measurements should be considered in a large energy range, from 100 keV to some MeV, for which data on energy straggling and multiple scattering in thin materials are missing.

At higher statistics, the accuracy of the realistic simulations performed until now using different tools, mainly SIMION and GEANT4, is not yet sufficient to reproduce properly the whole set of distributions measured in the experiments. This analysis has revealed that the ion cloud simulation has to be improved. New simulation tools, using GPU technologies, are thus developed, aiming to describe the physical processes involved during the ion confinement in the trap (cooling, space charge effects, interaction in the RF field) in the most realistic manner [22]. The statistical precision expected in the three cases (^{6}He, ^{35}Ar, ^{19}Ne) is given in Table 2, according to N_{coinc} (Table 1) and the precision obtained in the first experiment (2). For ^{35}Ar and ^{19}Ne, the SM values for $a_{\beta\nu}$ given in the table were calculated following Ref. [8]. The mixing ratios were deduced from the ratios between the mirror Ft-values and the $Ft(0^+ \rightarrow 0^+)$, including radiative corrections, but the extraction of the values of $a_{\beta\nu}$ does not include radiative corrections nor recoil effects. The current results are also given for comparison.

4 Conclusion and outlook

In addition to the RI charge state distributions induced by the shakeoff process, LPCTrap has provided data with sufficient statistics to significantly improve the current values of $a_{\beta\nu}$ in the allowed β decays of ^{6}He, ^{35}Ar and ^{19}Ne. If the realistic simulations achieved with the advanced tools in development at LPC Caen become sufficiently accurate to reproduce the whole set of measured distributions, the systematic uncertainties should remain at a level of precision comparable to the statistical uncertainties.

For ^{35}Ar, for example, the result would induce a significant gain (~ 1.7) on the V_{ud} precision deduced from the study of mirror decays [7]. This perspective motivates future measurements at GANIL in mirror decays, using new beams which are presently under development [25], such as ^{33}Cl and ^{37}K. In these two cases, precisions similar to the ^{35}Ar experiment are expected, considering an upgraded LPCTrap setup with increased detection efficiency, which is currently under investigation.

Acknowledgments The authors thank the LPC staff for their strong involvement in the LPCTrap project and the GANIL staff for the preparation of the beams.

Funding This work was supported in part by the Région Basse-Normandie, by the European networks EXOTRAP (contract ERBFMGECT980099), NIPNET (contract HPRI-CT-2001-50034) and EURONS/TRAPSPEC (contract R113-506065) and a PHC Tournesol (n° 31214UF).

Conflict of Interests The authors declare that they have no conflict of interest.

References

1. Severijns, N., Naviliat-Cuncic, O.: Phys. Scr. **T152**, 014018 (2013)
2. Naviliat-Cuncic, O., González-Alonso, M.: Ann. Phys. (Berlin) **525**, 600 (2013)
3. Severijns, N., Beck, M., Naviliat-Cuncic, O.: Rev. Mod. Phys. **78**, 991 (2006)
4. Fléchard, X., et al.: Phys. Rev. Lett. **101**, 212504 (2008)
5. Johnson, C.H., Pleasonton, F., Carlson, T.A.: Phys. Rev. **132**, 1149 (1963)
6. Glück, F.: Nucl. Phys. A **628**, 493 (1998)
7. Naviliat-Cuncic, O., Severijns, N.: Phys. Rev. Lett. **102**, 142302 (2009)
8. Severijns, N., et al.: Phys. Rev. C **78**, 055501 (2008)
9. Severijns, N., Naviliat-Cuncic, O.: Annu. Rev. Nucl. Part. Sci. **61**, 23 (2011)
10. Hardy, J.C., Towner, I.S.: Phys. Rev. C **79**, 055502 (2009)
11. Ban, G., et al.: Ann. Phys. (Berlin) **525**, 576 (2013)
12. Gorelov, A., et al.: Phys. Rev. Lett. **94**, 142501 (2005)
13. Vetter, P.A., et al.: Phys. Rev. C **77**, 035502 (2008)
14. Fléchard, X., et al.: J. Phys. G: Nucl. Part. Phys. **38**, 055101 (2011)
15. Li, G., et al.: Phys. Rev. Lett. **110**, 092502 (2013)
16. Van Gorp, S., et al.: Phys. Rev. C **90**, 025502 (2014)
17. Rodríguez, D., et al.: Nucl. Instrum. Methods Phys. Res. A **565**, 876 (2006)
18. Couratin, C., et al.: Phys. Rev. Lett. **108**, 243201 (2012)
19. Couratin, C., et al.: Phys. Rev. A **88**, 041403(R) (2013)
20. Fléchard, X., et al.: Hyperfine Interact. **199**, 21 (2011)
21. Velten, P.h., et al.: Hyperfine Interact. **199**, 29 (2011)
22. Fabian, X., et al.: These proceedings
23. Allen, J.S., et al.: Phys. Rev. **116**, 134 (1959)
24. Broussard, L.J., et al.: Phys. Rev. Lett. **112**, 212301 (2014)
25. Delahaye, P., et al.: AIP Conf. Proc. **1409**, 165 (2011)

Hyperfine Interact (2015) 236:9–18
DOI 10.1007/s10751-015-1189-x

Preparation of cold ions in strong magnetic field and its application to gas-phase NMR spectroscopy

K. Fuke[1,2] · Y. Ohshima[3] · M. Tona[4]

Published online: 14 May 2015

Abstract Nuclear Magnetic Resonance (NMR) technique is widely used as a powerful tool to study the physical and chemical properties of materials. However, this technique is limited to the materials in condensed phases. To extend this technique to the gas-phase molecular ions, we are developing a gas-phase NMR apparatus. In this note, we describe the basic principle of the NMR detection for molecular ions in the gas phase based on a Stern-Gerlach type experiment in a Penning trap and outline the apparatus under development. We also present the experimental procedures and the results on the formation and the manipulation of cold ions under a strong magnetic field, which are the key techniques to detect the NMR by the present method.

Keywords Nuclear magnetic resonance · Gas-phase molecular ions · Ultracold ion

1 Introduction

Nuclear magnetic resonance (NMR) technique is widely used for the physical and chemical analysis of various materials in liquid and solid phases [1]. Although this technique has been

Proceedings of the 6th International Conference on Trapped Charged Particles and Fundamental Physics (TCP 2014), Takamatsu, Japan, 1-5 December 2014

✉ K. Fuke
fuke@kobe-u.ac.jp

1   Institute for Molecular Science, Myodaiji, Okazaki, Japan

2   Graduate School of Science, Kobe University, Rokkodai, Nada-ku, Kobe, Japan

3   Department of Chemistry, Tokyo Institute of Technology, O-okayama, Meguro-ku, Tokyo, Japan

4   Ayabo Co. Fukukama, Anjo, Japan

well established, it requires a large amount of samples and also, sometimes, the purification and crystallization limit the structural analysis. In order to compensate these problems, recently, mass spectroscopic techniques are widely used in various research fields [2]. Since this method gives information only on the mass number, many research groups use it to reconstruct the structures of the parent molecular ions with an aid of the computer simulations for the fragment ions [3]. Especially, if the target molecular ion includes an isomer, which has the same mass number and the different geometrical structure, the assignment of geometrical structure becomes much difficult. Under these circumstances, a new extension of the NMR technique to the gas-phase molecular ions, which enables us to obtain rich information on the structure of the target ions with a mass-spectral sensitivity, becomes increasingly important in both fundamental and applied sciences. Historically, the NMR spectroscopy was first used for studying the magnetic moment of isolated atoms and molecules by Rabi and coworkers [4]. They succeeded in determining the magnetic moments by developing a molecular beam magnetic resonance technique based on the Stern-Gerlach experiment. Unfortunately, this technique utilized to neutral species is not applicable to the gas-phase ions, because the ions cannot be confined under the strong wedge-type magnetic field. Quite recently, Ulmer and his coworkers successfully reported a first NMR detection of a single trapped proton [5] with a continuous Stern-Gerlach experiment similar to that developed by Dehmelt [6]. However, this technique is limited to light nuclei, because they need a much stronger inhomogeneous field to detect a spin flip of the heavier nuclei.

In order to resolve the aforementioned problems on the structural analysis of the gas-phase molecular ions, we proposed a new method, called "magnetic resonance acceleration" technique, in our previous paper [7]. The outline of our detection method will be described in Experimental Section. Briefly, we adopt a long Penning-type trap as a NMR cell and place it in a bore of superconducting magnet, which has two homogeneous field regions with high and low magnetic fields and a strong gradient field in between [8]. To detect a weak NMR signal, two radio frequency (RF) coils are installed next to the trapping electrodes in the NMR cell to flip the nuclear spin of the ions synchronized with the shuttle motions of the ions. This configuration allows multiple interaction between the magnetic moment of the ions with the gradient magnetic field and realizes a continuous Stern-Gerlach experiment for the trapped ions.

As in the Rabi's method, the key techniques of the present method to detect the NMR signal lie in the formation of cold molecular ions and their manipulation. Until now, various methods have been developed for cooling neutral atoms and molecules [9]. For atomic and diatomic ions, a variety of sophisticated techniques have been developed, such as the laser based cooling methods [10]. A sympathetic cooling is an interesting technique to cool complex molecular ions [11, 12]. However, it is difficult to combine these techniques to the device for the NMR detection, which requires an extremely homogeneous magnetic field region for conducting a high-resolution magnetic resonance experiment. Although ions are easy to manipulate with electric and magnetic fields, slow ions are very sensitive to stray fields and space charge. Especially, when the trapping space is large as in the case of our apparatus, it becomes very difficult to prepare cold molecular ions and manipulate them [10].

In the present study, we newly design and construct the ion source and the NMR cell containing the RF coils to reduce the effect of stray fields. As mentioned above, the present method requires the ultracold ion packets with a slow velocity and an extremely narrow velocity distribution width. In order to develop the cooling technique of molecular ions, we explore the methods to decelerate the ion packets and bunch them under the strong magnetic

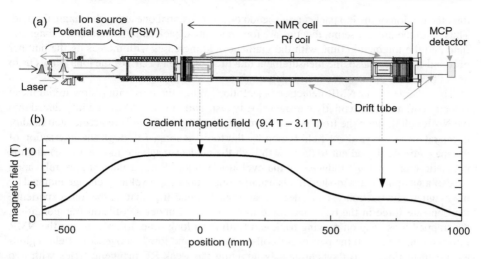

Fig. 1 a Experimental setup for the potential switch and the NMR cell. A small hole on the drift tube of potential switch is a window for introducing the photoionization laser. **b** Magnetic field distribution of the superconductive magnet. The scales of these figures are the same and, the arrows indicate the center of the high and low homogeneous magnetic field regions

field by using a potential switch and a velocity selection. Here, we outline the principle of the gas-phase NMR detection. We further describe the features of the newly developed apparatus and discuss on the experimental results on the preparation and manipulation of cold molecular ions.

2 Experimental method

2.1 Principle of NMR detection

A detailed description of our method was published previously [7] and so only the essential details of the experiment scheme are summarized below. The present method is a new extension of the Rabi's technique for neutral molecules. In their technique, they detected a deviation of the radial distribution of molecular beam [4], which is induced by the interaction of nuclear magnetic moment with a strongly gradient magnetic field generated by a wage-type magnet aligned in the beam direction. Because the molecular ion beams cannot be confined under the gradient magnetic field, we newly design a Penning trap for the NMR cell to conduct the Stern-Gerlach type experiment. Figure 1a shows a scheme of our experimental setup, which is installed in the bore of the superconducting magnet. The magnet has two homogeneous field regions with high ($B_H = 9.4$ T) and low magnetic ($B_L = 3.1$ T) fields as shown in Fig. 1b [7, 8], where a magnetic resonance excitation of the ions is performed with two different oscillatory fields. The above two regions are connected with a steep gradient field region, where the ions experience the magnetic force in the axial direction. The front and back gates are placed at near the 9.4 T and 3.1 T regions, respectively, to store the ions in the z direction. When the ions are injected into the cell, they are confined from escaping in the xy plane under the strong axial magnetic field by virtue of a cyclotron motion. As described in the previous paper [7], the features of the NMR cell allow us to treat the motion of the ion packet as a pseudo one dimensional one along the magnetic field. Thus, with the present setup, the ions are forced to move back and forth along the magnetic field and the interaction time with the gradient field increases with increasing the number of round-trips. And also, the measurement of a time-of-flight (TOF) of the ions allows us to detect the effect of magnetic interaction on the axial velocity.

The scheme of our NMR detection is as follow. When the ions with a slow initial mean velocity (v_0) and a sufficiently narrow velocity distribution width (Δv_0) are introduced into the NMR cell through the front gate, they move along the z axis and are accelerated (and/or decelerated) at the gradient-field region by the force generated through the interaction of the magnetic field gradient (dB/dz) with both the nuclear magnetic moment (μ_N) and the magnetic moment (μ_C) induced by the cyclotron motion. Since a nuclear spin relaxation time of a gas-phase closed-shell ion is usually longer than 1 s, a velocity change induced by the magnetic interaction is cancelled out after each round-trip, that is, the ions experience the opposite force in the back and force motions. Thus, under a collision free condition, the trapped ions keep on moving back and forth for a long time. In order to get the NMR information, we install the pair of RF coils in the high and low homogeneous field regions as shown in Fig. 1a. By synchronously applying the weak RF magnetic fields with two resonance frequencies so called a π-pulse to the ions at each time when they pass through the coils, the nuclear spins flip if a resonance condition is fulfilled. With this scheme, the translational velocities of the ions in the different nuclear spin states increase or decrease continuously with increasing the number of round trips. After an appropriate number of round-trips (N), the ions with the different nuclear spin states are spatially separated in the z direction ("spin polarization"). This change can be observed by measuring the TOF profile of the ion packet released from the NMR cell with a microchannel plate (MCP) detector. As described in the previous paper [7], the measurements of the time profiles as a function of the RF frequency allow us to determine the nuclear magnetic moment and a Larmor frequency of the proton (ω) in the molecular ions. It should be noted that the frequencies of two coils installed at the high and low magnetic fields should satisfy a condition that $\omega_H/\omega_L = B_H/B_L$, where ω_H and ω_L are the Larmor frequencies of proton at two homogeneous fields; about 400.28 and 132.17 MHz for proton nuclei, respectively.

The present method enables us to detect the spatial spin polarization in the axial direction by narrowing the translational velocity distribution, that is, by cooling the translational temperature of the molecular beam. The detection sensitivity of the spin polarization is mainly determined by the initial conditions of the translational motions of the sample ions; mean velocity (v_0), velocity distribution width $(\Delta v_0;$ full width half maximum), and temporal width (τ_0). Under the gradient magnetic field, the force (F) acts on nuclear spin is given by $F_z = \mu_N dB(z)/dz$, where μ_N is the z component of the magnetic moment $(\mu_N = \gamma (h/2\pi) m_I)$; m_I is a nuclear spin quantum number $(= 1/2)$. A simple calculation gives the velocity increase (Δv) as

$$\Delta v = (\mu_N/Mv_0)(B_H - B_L)(2N + 1), \tag{1}$$

where M is the mass of the ion. μ_N is 1.41×10^{-26} J T^{-1} for proton. In the case of molecular ion having n_p protons with the same chemical environment (Larmor frequency), Δv for the nuclear spin state having the highest total quantum number is given by multiplying n_p to (1). An estimation of magnetic resonance acceleration was described in the previous paper [7]. Similar calculations for the NH$_3^+$ ions predict the number of round trips (N) as $N \geq 4$ to detect the spin polarization with the initial conditions as $v_0 = 100$ m/s (1.8 meV), $\Delta v_0 = 0.4$ m/s, and $\tau_0 = 50$ μs. These estimations indicate that the present method requires

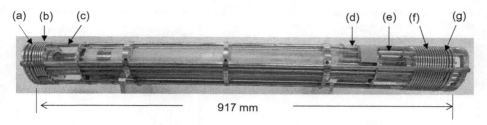

Fig. 2 Photograph of the NMR cell: (**a, g**) mesh electrode pairs, (**b, f**) cylindrical electrodes, (**c, e**) RF coils, (**d**) coil tuning unit consisting of non-magnetic trimer capacitors

the ions with very slow velocity with a narrow velocity width to observe the spatially well resolved spin polarization. Thus, as in the case of the Rabi's experiment, it is important to establish the technique to cool the translational temperature of the ions by narrowing the distribution width of the longitudinal velocity.

2.2 Experimental setup

In the present experiment, we adopt the photoionization of a supersonically cooled molecular beam to precool the ions. The ion source system consists of two differentially evacuated chambers separated by a 2-mm-diameter skimmer, which are evacuated by turbo molecular pumps. A pulsed valve (Jordan, PSV) is placed 20 mm upstream of the skimmer. The molecular beam is collimated further with another 5-mm-diameter skimmer installed at 800 mm downstream from the pulsed valve and is introduced into a third chamber, which is evacuated with a cryopump down to below 10^{-7} Pa. Figure 1 shows a schematic drawing of the experimental setup for the photoionization source working as a potential switch and the NMR cell. In the present study, both the ion source and the NMR cell are newly constructed to improve the cooling condition of the ions. The ion source, which consists of a 400 -mm length drift tube and cylindrical electrodes, is placed at the upstream of the NMR cell. The photoionization laser is crossed with the neutral beam through a hole located at the wall of the drift tube as shown in Fig. 1a. To facilitate the deceleration and the compression of the ion packet, the potential switch consisting of 17 cylindrical electrodes are divided into two parts. The first part is used as a buncher, while the second part is used to generate a potential slope for the deceleration. The drift tube and cylindrical electrodes are biased through a series chain of non-magnetic resistors.

The NMR cell shown in Figs. 1a and 2 is the Penning-type trap with an effective length of 744 mm. The front and back gates of the trap are made of five cylindrical electrodes and store the ion packet for a long time without a loss of the ions. On the other hand, the mesh electrode pairs installed both at the upstream of the front gate and at the downstream of the back gate work as a velocity selector for slicing the velocity distribution, as described in the next section. A pair of the home-made saddle-type RF coils are mounted just at the downstream and the upstream of the front and back gates, respectively. These coils are placed exactly at the center of the high (9.4 T) and low (3.1 T) homogeneous magnetic field regions as shown in Fig. 1 and are aligned parallel to the magnetic axis so as to generate the RF magnetic fields in the direction perpendicular to the ion beam. All electrodes and the RF coils are made of Au/Ag plated copper. The circuit including the RF coils is tuned by two closely-located non-magnetic capacitors (see (d) in Fig. 2); they are adjusted from the

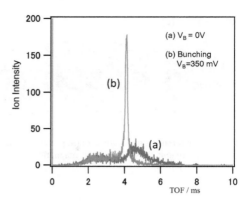

Fig. 3 Time-of-flight profiles of the NH$_3^+$ ions without (**a**) and with the bunching process (**b**). The bunching voltage is $V_B = 350$ mV

outside of the vacuum. The basic performance of these coils as the NMR probe is examined by measuring a proton NMR spectrum of a liquid-water sample tube placed in the coil.

As described in the next section, we have to prepare the ions with a very small kinetic energy. To manipulate the ions in the NMR cell, a baking process is indispensable; in fact, without the baking, the ions with the kinetic energy of less than 100 meV do not pass through the NMR cell. A possible cause of this defect is considered to be the mesh electrodes installed in the NMR cell, where the ions come closest to metal surfaces and may be influenced by a patch effect. In the present study, we also newly developed a non-magnetic baking system to heat the whole NMR cell up to 400 K.

3 Formation and manipulation of cold ion packet

In order to prepare the ions with a slow velocity of a very narrow velocity distribution width, the ions are cooled by three steps. We start with a supersonic molecular beam to precool the neutral molecules. The photoionization of the molecular beam enable us to form the cooled ion packet as cold as less than 10 K [13]. The molecular ions produced by these procedures are internally cold and have a narrow relative velocity distribution. However, the mean velocity of the ions in the laboratory frame is relatively high, and it depends on the temperature and the pressure of the source as well as the mass of the carrier gas employed [13]. The present experiment requires much lower mean velocity and velocity distribution width. Thus, in the second step, the ions produced are cooled by the bunching and deceleration using the potential switch. And then, the ions are transferred to the NMR cell and cooled further by a velocity selection procedure, in which the velocity distribution is reduced by cutting out a part of the ion packet with the mesh electrode pair. In the following sections, we describe the details of these experimental results for the formation and manipulation of the cold ion packet.

3.1 Deceleration and bunching of ion packet.

A mixture gas of NH$_3$ with Ar is expanded through the pulsed valve with a nozzle diameter of 0.5 mm. To facilitate the formation of monomer NH$_3^+$ ions, we use an amplified output of a picosecond laser pulse at 193 nm (20 ps, 3 mJ) as the photoionization light source by using an ArF excimer laser. A seed laser light at 193 nm is generated by mixing the fundamental of a picosecond YAG laser (EKSPLA PL2143) with the second harmonic of an optical parametric output at 236.6 nm. Figure 3 displays the TOF profile of the ions produced in

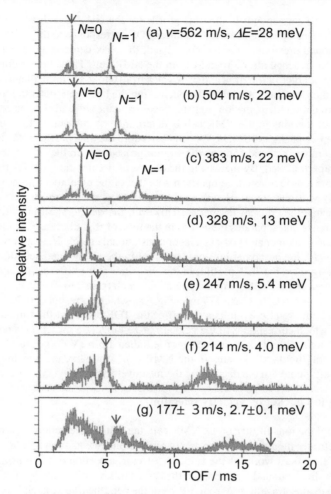

Fig. 4 Time-of-flight profiles of the NH_3^+ ions at various deceleration potentials. The mean velocities after the deceleration are indicated with the kinetic energy. The peaks with the arrows (noted as $N = 0$) are those directly reached to the MCP without reflection at the back gate, while those at the longer TOF (note as $N = 1$) correspond to the ion packets reached after one round-trip in the NMR cell. The second arrow at 17 ms in Fig. 4g indicates the transmission limit of the decelerated ions

the ion source. The curve (a) corresponds to the ion signals without the bunching. The ion signals consist of two broad peaks, in which the slower peak corresponds to the NH_3^+ ions. The weaker peak reached earlier to the detector is assigned to the aggregate ions such as $NH_4^+(NH_3)_n$. In the bunching process, the pulsed voltage is applied to the buncher with a suitable delay time after the photoionization laser. The curve (b) in Fig. 3 shows the result of the bunching with the pulsed voltage of 350 mV. By this procedure, the monomer ions are selectively bunched and their velocity distribution becomes very narrow without changing the mean velocity significantly.

These ions are decelerated by applying a continuous voltage to the rest of the cylindrical electrodes, which work as the potential switch. Figure 4a shows the TOF profiles for the NH_3^+ ions obtained by setting the voltage of the potential switch to 0 V. Here, the peak denoted as $N = 0$ corresponds to the time profile for the NH_3^+ ions directly reached to the MCP, while the weak peak appeared at a shorter TOF corresponds to those for the aggregate ions as mentioned previously. On the other hand, the peak denoted as $N = 1$ corresponds to the NH_3^+ ions reached after a round-trip in the NMR cell. The latter profile is obtain by changing the potentials of the front and back gates from 0 to 0.3 V immediately after the ions pass through the front gate. And then, by decreasing the back-gate potential to 0 V after an elapsed time of a round-trip, the ion packet escapes from the cell and is detected by the MCP. The velocity of the ions in the NMR cell is determined by measuring the time difference between these two peaks as 562 m/s. We can also record the time profile of the ion packet for any desired number of round-trips by the similar procedures. In the figure, these two time profiles are superimposed. By increasing the degree of deceleration, the relative intensities of the aggregate ions increase in comparison with that of the monomer ions as seen in Fig. 4. These ions fly with the same velocity as the monomer ions and are detected at the same arrival time when they do not decelerate. However, the deceleration potential causes the velocity difference and separates these ions in the time of flight because the cluster ions are heavier than the monomer and possess a lager translational energy. Hereafter we concentrate on the analysis for the monomer ions only to simplify the following discussion.

Figures 4a-4g show the TOF profiles of the NH_3^+ ions at the different degree of deceleration. With decreasing the voltage of the potential switch from 0 to -0.2 V, the mean velocity of the ions slows down to less than 177 m/s (Fig. 4g), which correspond to the kinetic energy of 2.8 meV. Figure 4g also exhibits a cut off of the TOF profile at the time of longer than about 17 ms, which corresponds to the velocity of 134 m/s (1.5 meV). This result indicates that the NH_3^+ ions with the kinetic energy of less than 1.5 meV do not reach to the detector due to a stray field still remains in the NMR cell. And also, the bunching of the ions becomes not efficient with the decreasing the mean velocity.

3.2 Cooling ion packet by velocity selection

In order to cool the ions further in the NMR cell, the velocity selection is carried out using the mesh electrodes mounted at the upstream of the front gate as shown in (a) of Fig. 2. The pair of mesh electrodes works as the velocity selector; a portion of the ions in the velocity distribution is sliced out and is restored in the cell as follow. When the ion packet comes up to the front gate after travelling N round-trips in the cell, the trap voltage is reduced to 0 V and, and at the same time, a positive pulse with a suitable duration is applied to the mesh installed at a upstream side. With these procedures, only the center portion of the ion packet is restored in the NMR cell and the rest of the ions escapes from the confinement. As an example, the time profiles up to $N = 5$ for the ion packet injected with $v_0 = 274$ m/s are shown in Fig. 5a, while, those sliced at $N = 2$ are shown in Fig. 5b. The pulse width and the amplitude applied to the mesh electrode are set at 200 μs and 0.5 V, respectively. As seen in Fig. 5a, the width of the ion packet becomes broad with increasing the number of round trips. These changes are a thermal broadening due to the initial velocity distribution width (Δv_0). As discussed in the previous paper [7], Δv_0 is obtained from the time profiles by the $\Delta t - t_C$ plot, where t_C and Δt are the center and the width of each time-profile peak, computed by fitting the observed peaks with assuming a Gaussian function. Without the slicing, the slope of the plot gives the velocity distribution width as $\Delta v_0 = 20$ m/s. After the slicing, Δv is calculated to be 2.4 m/s FWHM. Thus, the velocity distribution width is reduced by about one tenth with the velocity selection using the slicing technique. By assuming the velocity distribution of the ions as a one-dimensional Maxwell-Boltzmann type, Δv can be related to the translational temperature [7]. The width before the slicing (20 m/s) corresponds to

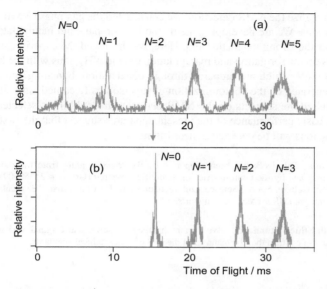

Fig. 5 Time-of-flight profiles of NH_3^+ ion packet (**a**) after injecting the NMR cell and (**b**) after slicing at the second round-trips ($N = 2$) are shown. The pulse duration for slicing is 200 μs. The peaks for $N = 0 - 3$ in Fig. 5b are those traveling in the NMR cell after the slicing for ($N+2$)th round trips

the translational temperature as 0.16 K, and after the velocity selection, the temperature successfully reaches down to 2.2 mK. As described in Section 2.1, however, the velocity distribution width must be less than 1 m/s FWHM to facilitate the NMR detection for NH_3^+. Thus we still need to improve the cooling technique for molecular ions.

During the course of these studies, we found a very effective velocity selection procedure for p-xylene ions as a sample. In this experiment, a negative offset voltage of a few 100 mV was applied to the front gate in advance at the time of the velocity selection. A preliminary result indicates that the velocity distribution width was reduced further down to about 0.4 m/s FWHM (0.3 mK) . In order to clarify the mechanism to generate the extremely narrow velocity distribution width, we carried out a simulation on the velocity selection process using a software, Simion. The calculated results indicate that the negative offset voltage on the front gate induces a dispersion compensation of the distribution of the velocity-selected ions, which is similar to the bunching process. In order to succeed in detecting the first NMR signals of the gas-phase molecular ions, we are currently conducting the experiments including the improvement of the cooling technique based on these results.

4 Summary

NMR technique is a powerful tool to study the physical and chemical properties of materials in wide area. However, this technique is limited to the materials in condensed phase because of its very low sensitivity. In order to break this situation and to overcome the sensitivity problem, we proposed the new principle to detect the NMR of gas-phase molecular ions based on the Stern-Gerlach type experiment in the Penning trap. In this method, the ultra-cold molecular ions are introduced in the trap and their magnetic moments are probed by observing the modulation of their TOFs induced by the RF magnetic excitation at both ends of the trap. To realize the NMR detection, we constructed the gas-phase NMR apparatus for mass selected ions. We are developing the methods to prepare and manipulate cold molecular ions under the strong magnetic field. Here, we discussed the experimental techniques and the results on the formation and manipulation of cold NH_3^+ ions with the kinetic energy of less than 3 meV, which are prepared through deceleration, bunching and slicing of the ion packets generated by the photoionization of supersonically cooled ammonia molecules. We also discussed the subjects on the NMR detection for cold mass-selected ions. These results on the basic performance of the present apparatus suggest that the first NMR signal for polyatomic ions will be detected in near future.

Acknowledgments This work has been supported by the research grant from the Japan Science and Technology Agency and by the Grant-in-Aid for Scientific Research (Grants # 24350009) from the Ministry of Education, Culture, Sports, Science, and Technology (MEXT) of Japan. We thank the Equipment Development Center of IMS for technical supports.

Compliance with Ethical Standards We warrant that the manuscript is an original work and has not been published before. We confirm that the manuscript meets the highest ethical standards.

References

1. Fukushima, E., Roeder, S.B.: Experimental Pulse NMR. Westview Press (1993)
2. Sinz, A.: Mass Spectrom. Rev. **25**, 663 (2006)
3. Syrstad, E.A., Turecek, F.: J. Am. Soc. Mass Spectrom. **16**, 208 (2005)
4. Rabi, I.I., Millman, S., Kusch, P., Zacharias, J.R.: Phys. Rev. **55**, 526 (1939)
5. Ulmer, S., Rodegheri, C.C., Blaum, K., Kracke, H., Mooser, A., Qunt, W., Walz, J.: Phys. Rev. Lett. **106**, 253001-1 (2011)
6. Van Dyck, R.S., Schwinberg, P.B., Dehmelt, H.: Phys. Rev. Lett. **59**, 26 (1987)
7. Fuke, K., Tona, M., Fujihara, Ishikawa, H.: Rev. Sci. Instrum. **83**, 085106-1-8 (2012)
8. Kominato, K., Takeda, M., Minami, I., Hirose, R., Ozaki, O., Ohta, H., Tou, H., Ishikawa, H., Sakurai, M., Fuke, K.: IEEE Trans. Appl. Supercond. **20**, 736 (2010)
9. Schnell, M., Meijer, G.: Angew. Chem. Int. Ed. **48**, 6010 (2009)
10. Gerlich, D.: The production and study of ultra-cold molecular ions. In: Smith, I.W.M. (ed.) Low temperatures and cold molecules, pp. 295–343. Imperial College Press (2008)
11. Larson, D.J., Bergquist, J.J., Itano, W.M., Wineland, D.J.: Phys. Rev. Lett. **57**, 70 (1986)
12. Ostendorf, A., Zhang, C.B., Wilson, M.A., Offenberg, D., Roth, B., Schiller, S.: Phys. Rev. Lett. **97**, 243005 (2006)
13. Anderson, J.B., Fenn, J.B.: Phys. Flu. **8**, 780 (1965)

Hyperfine Interact (2015) 236:19–27
DOI 10.1007/s10751-015-1194-0

Electron attachment to anionic clusters in ion traps

Franklin Martinez[1] · Steffi Bandelow[2] · Gerrit Marx[2] ·
Lutz Schweikhard[2] · Albert Vass[2]

Published online: 28 May 2015

Abstract Ion traps are versatile tools for the investigation of gas-phase cluster ions, allow-ing, e.g., cluster-size selection and extended reaction times. Taking advantage of their particular storage capability of simultaneous trapping of electrons and clusters, Penning traps have been applied for the production of clusters with high negative charge states. Recently, linear radio-frequency quadrupole traps have been demonstrated to be another candidate to produce polyanionic clusters. Operation with rectangular, rather than harmonic, radio-frequency voltages provides field-free time slots for unhindered electron passage through the trap. Several aspects of electron-attachment techniques by means of Penning and radio-frequency traps are addressed and recent experimental results are presented.

Keywords Polyanions · Metal clusters · Penning trap · Radio-frequency trap · Digital ion trap

1 Introduction

Multiply negatively charged atoms, molecules and clusters in the gas phase have been sub-ject to experimental and theoretical investigation for a long time [1–3]. For example, doubly charged anions of carbon and metal clusters were produced either in ion sources by sputter-ing [4], laser ablation [5–7], and electrospray ionization [8], or by electron-transfer reactions [9, 10], and in crossed-beam experiments [11]. While also tri-anionic lead clusters were

---

Proceedings of the 6th International Conference on Trapped Charged Particles and Fundamental Physics (TCP 2014), Takamatsu, Japan, 1–5 December 2014.

✉ Franklin Martinez
franklin.martinez@uni-rostock.de

[1] Institute of Physics, University of Rostock, Rostock, Germany

[2] Institute of Physics, Ernst-Moritz-Arndt University, Greifswald, Germany

                     Springer

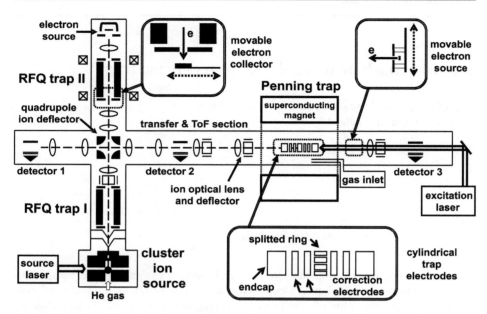

Fig. 1 Schematic of the present ClusterTrap setup, with cluster ion source, RFQ traps and Penning trap. Details of the electron collector at the RFQ trap II, of the cylindrical Penning trap, and of the electron source at the Penning trap are shown in insets

observed by laser ablation [7], even higher charge states have been produced of gold and of aluminum clusters by their simultaneous storage with electrons inside a Penning trap [12–18]. Recently, production of di- and trianionic gold clusters in a linear radio-frequency quadrupole (RFQ) trap has been demonstrated by exposing stored monoanions to an electron beam [16, 17]. In the present report, methods for electron attachment to trapped cluster anions are characterized for both Penning and RFQ traps. Gold cluster dianions have been produced in an RFQ trap that has been operated in a digital-ion-trap (DIT) mode [19–22], including field free time-periods [23].

2 Experimental setup

The experiments have been performed at the ClusterTrap, a setup designed for production and investigation of gas-phase cluster ions [17, 24–26]. For the present studies, metal clusters have been produced by laser irradiation of a metal wire and subsequent expansion of the vapor into vacuum by means of a helium gas pulse [27]. By variation of the laser and gas-pulse parameters, the cluster distribution is shifted with respect to the cluster size, i.e. the number of atoms. Of the neutral and singly charged clusters produced in the source, the mono-anionic ones are captured and accumulated in a linear radio-frequency quadrupole ion trap (RFQ trap I in Fig. 1). After ejection from this trap, they are guided by an electrostatic quadrupole deflector and other ion-optical elements into a 12-T cylindrical Penning trap (inset in Fig. 1) [17, 24]. During in-flight capture and storage, cluster ions are mass-over-charge selected and exposed to interaction steps of interest. The latter involve an electron beam from a movable, heated rhenium filament (inset in Fig. 1), and a laser beam with pulse durations in the nanosecond range. Charged reaction products remain stored and are subsequently analyzed by time-of-flight (ToF) mass spectrometry, for the present study by ejection from the Penning trap and drift to detector 1 (Fig. 1).

In an alternative experimental sequence, cluster monoanions accumulated in RFQ trap I are transferred into another linear ion trap (RFQ trap II, Fig. 1), which is particularly designed and equipped for the investigation of polyanion production in a radio-frequency trap [17]. The heated filament of a small, cut-open halogen lamp is located behind the RFQ trap II and provides an electron beam along the trap axis, which can be monitored by a movable electron collector plate (inset in Fig. 1). For unhindered passage of electrons through the RFQ trap II, it is operated as a three-state digital ion trap [23], i.e. the radio-frequency voltage is realized by well-tailored, fast (kHz) switching between three electric potentials, instead of the usual harmonic voltage signals. One of the three potential steps corresponds to zero-volt potential differences between the RFQ rods, i.e. few-microsecond intervals of a field-free trap volume [23]. The passage of the electron beam is supported by a low magnetic field of about 20 mT, orientated parallel to the trap axis. It is provided by two induction coils located outside the vacuum vessel (Fig. 1).

3 Electron attachment in RFQ trap II

Production of polyanionic metal clusters in a linear radio-frequency trap is realized by attachment of electrons to stored cluster monoanions. For this purpose, an electron beam is guided along the trap axis during the field-free time slots of the digital radio-frequency trapping voltage, i.e. the clusters are exposed to a sequence of microsecond electron-beam pulses.

From the many collisions between cluster anions and electrons, only some will result in the attachment of an electron, i.e. in the increase of the negative charge state of the cluster. However, this will be the case only if the cluster is large enough to accept and (meta-)stably bind another electron [3, 13, 18, 28]. The relative yield of polyanionic clusters is mainly determined by the electron energy during the interaction process. For very high electron energies, a collision might result in "kicking out" an electron from the anionic cluster, i.e. electron loss, rather than electron attachment. However, for attachment, the approaching electron has to overcome the repulsing Coulomb potential of the already negatively charged cluster, i.e. it has to exceed a minimum energy value. The Coulomb-barrier height, and hence the required electron energy, increases with the negative charge state of the cluster. Thus, the energy distribution of the incident electron beam is crucial not only for the yield of polyanionic clusters, but also for the reachable charge state. Note, that a more technical criterion, due to the trapping conditions, may limit the reachable charge state even stronger: while some higher charge state might be produced by electron attachment, the mass-over-charge ratio of the resulting cluster polyanion may prevent it from being further trapped, i.e. in the present experimental setup those clusters are lost from the RFQ-trap before being detected. This is different from the Penning trap where there is no lower storage limit of the mass-over-charge range.

The energy of the attaching electrons is to first order approximated by the difference between the filament floating potential and the offset potential of the ion trap, assuming the stored clusters are cooled into the axial trapping well, e.g. by application of buffer gas. Then, the mean energy is easily controlled by varying the filament float potential. However, the width of the electron-energy distribution depends on the type of electron source. In the present case of a resistively heated filament, it is determined by the voltage drop between both ends of the filament, caused by the heating current. Figure 2 shows the emitted current

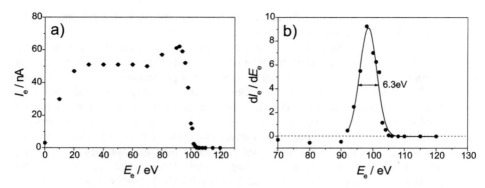

Fig. 2 **a** Electron current I_e emitted from a heated filament source as a function of the electron energy E_e. **b** The derivative dI_e/dE_e reflects the distribution of the electron energy (*symbols*), and is fitted with a Gaussian function (*solid line*) yielding a mean energy of 98.5 eV and a FWHM of 6.3 eV

of such a type of filament electron source as a function of the electron energy. The latter is determined by a blocking potential applied to a grid in front of an electron collector. (Note, that this analysis has been realized in a separate setup, i.e. without the RFQ trap II.) From the derivative of the measured current, the full width at half maximum of the energy distribution is determined to be about 6 eV, while the energy width at the base level reflects the voltage drop along the filament of about 10 eV.

It is planned for future experiments to determine the Coulomb potential height for a given cluster size and charge state by measuring the yield of ions with the next higher charge state as a function of electron energy. However, such a task requires energy distributions in the few meV range to resolve, e.g., the potential heights of clusters with the same charge state, but with different sizes. Respective electron-energy distributions may be obtained either by more suitable types of electron sources, e.g. indirectly heated ones, or by combining the present source with an electron energy selector with meV-resolving power.

In particular low-energy electrons are easily diverted from their initial direction of motion by weak electric fields. However, at the same time they are well guided by magnetic fields low enough to keep the motion of atomic ions, let alone of cluster ions, undisturbed. The magnetic field superimposed on the RFQ trap II [17] has been analyzed prior to installation of the induction coils at the ClusterTrap setup. The axial magnetic field inside the vacuum vessel (but under atmospheric pressure) has been measured by use of a Hall probe (Fig. 3a). The two circular coils produce maxima of the magnetic field strength close to the positions where the electrons enter and leave the trapping volume through holes in the endcaps (EC1 and EC2 in Fig. 3a). Between the endcaps the field strength drops to about 80 % of the maximum, still sufficient to guide the electrons.

After installation of the coils at the setup, the electron current through the trap has been monitored as a function of magnetic field strength B (Fig. 3b) by means of the movable collector located between the electrostatic ion deflector and the RFQ trap II (Fig. 1). While no electron current is detected behind the trap without any magnetic field, about 30 μA are measured for $B = 20$ mT.

In recent experiments, electron attachment in the RFQ trap II operated in the 2-state DIT mode (Fig. 4a), i.e. without field-free time intervals in the RF-signal, had been realized. As a result, production of gold cluster di- and trianions [16], as well as combined polyanion production with attachment first in the RFQ and then in the Penning trap, has been demonstrated [17].

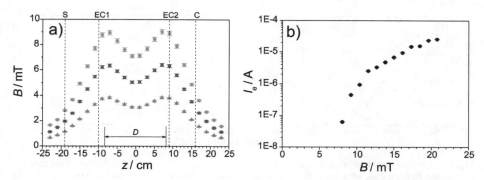

Fig. 3 a Magnetic field inductivity B as a function of the z-position along the axis of RFQ trap II for different coil currents (1.4A/24.5V red circles; 1.0A/17.4V blue squares; 0.6A/10.4V green triangles). The coil separation $D=16.5$ mm and the positions of trap endcaps (EC), electron source (S) and electron collector (C) are indicated. **b** Current on the electron collector (C) at the RFQ trap II (without application of an RF-field) as a function of the applied magnetic field

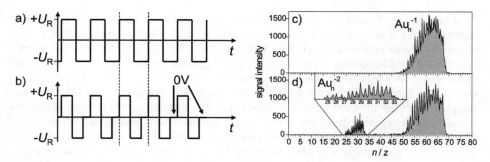

Fig. 4 a-b Schematic of the rectangular radio-frequency potential of a 2-state (**a**), and of a 3-state digital ion trap (**b**) with a duty cycle of 0.4 [23], i.e. 20 % of the period are at 0-V level. **c-d** Mass-spectra of gold clusters after accumulation in RFQ trap I, transfer to RFQ trap II and irradiation with an electron-beam, transfer to the Penning trap and ejection from there towards detector 1. **c** Reference spectrum of monoanions Au_n^{-1} ($n = 50$ to 70) stored for 2.6 s in RFQ trap II operated in the 3-state DIT-mode ($f_R = 65$ kHz, $U_R = 70$ V), without application of an electron-beam. **d** Dianions Au_n^{-2} produced after exposing the gold monoanions to 17-eV electron pulses of 2 μs duration over a period of 20,000 RF-cycles

Production of dianionic gold clusters in the RFQ trap II operated in the 3-state DIT mode (Fig. 4b) has been realized for the first time. Gold cluster monoanions, Au_n^{-1}, produced by the cluster source were trapped in the RFQ trap II for 2.6 s (Fig. 4c). Application of electron pulses of 2 μs duration over about 20,000 RF-cycles (i.e. total interaction time of 40 ms) results in the appearance of dianionic gold clusters, Au_n^{-2}, in the spectrum (Fig. 4d). The inset shows signals at half integers of the size-over-charge ratio n/z due to the negative charge state $z = 2$. While the previous experiments with the 2-state DIT required an electron energy of several electronvolts to traverse the RFQ trap, the application of the 3-state DIT with intermediate field-free time periods prepares experiments for electron attachment with sub-electronvolt energies.

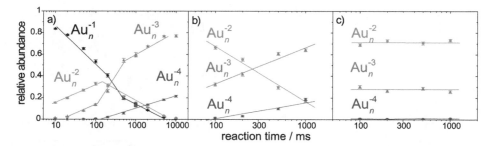

Fig. 5 **a** Relative abundance of gold anions Au_n^{-z}, $n = 127 \pm 15$, $z = 1 - 4$, as a function of the reaction time in the Penning trap, after application of an electron beam for 20 ms. **b-c** Relative abundance of gold polyanions, Au_n^{-z}, $n = 135 \pm 15$, $z = 2 - 4$, as a function of storage time, after application of an electron beam for 20 ms, without (**b**) and with removal of electrons (**c**) after a reaction time of 100 ms by pulsing the endcap electrodes for 1 μs. The *lines* are included to guide the eyes

4 Electron attachment in a Penning trap

First production of gas-phase polyanionic metal clusters in a 5-T Penning ion trap was reported for di- and tri-anions of gold, silver and titan [3, 12, 13, 29]. Later, di-, tri-, tetra-, and penta-anions of aluminum clusters were observed [14, 15, 30, 31]. With the recent upgrade of the ClusterTrap setup to a 12-T Penning trap, higher charge states for gold clusters have been reported, reaching from tetra- up to hexa-anions [17, 18]. Electron attachment in the Penning trap is realized by the electron-bath technique, i.e. the simultaneous trapping of cluster monoanions and electrons [12]. In short, during the storage of cluster monoanions, an electron beam and argon gas are simultaneously pulsed into the trap volume. Low-energy secondary electrons are produced by electron-impact ionization of argon gas atoms and remain trapped, while the argon cations and the high-energy primary electrons leave the trap. During a reaction time of typically 1 s polyanionic clusters are formed and remain stored until they are extracted and analyzed.

Besides the cluster-size criterion [3, 13, 18, 28], the maximum charge state and the relative yields of the polyanionic species are to some extent controlled by the trapping potential depth, which limits the energy of trapped electrons [14, 31–34], and by the reaction time [31, 34] as shown for the case of gold clusters (Fig. 5a). While the distribution is dominated by dianions and non-converted monoanions after 10 ms, those species disappear after a few seconds in favor of tri- and tetra-anions. For application of further experimental steps on a given charge-state distribution, the reaction time is intentionally terminated by removing the stored electrons from the trap without affecting the stored cluster anions. It is realized by a 1-μs pulsing of the endcap electrodes (Fig. 5b-c), a technique known as suspended trapping [35, 36]. Thus, lifetime measurements on meta-stable polyanionic clusters may become feasible.

In the present setup, the electron source, made of a resistively heated stripe of rhenium foil, is mounted on a horizontally movable support and is located just outside the 12-T superconducting magnet, i.e. in the region of a strong magnetic-field gradient. Thus, electrons are guided close to the axis of the Penning trap, even if they are emitted from an off-axis position, as monitored by the production of polyanionic gold clusters (Fig. 6a). Thus, production and subsequent laser excitation of polyanionic clusters [34, 37–39] can be combined in an experimental sequence, where the electron beam and the laser beam enter the trap from the

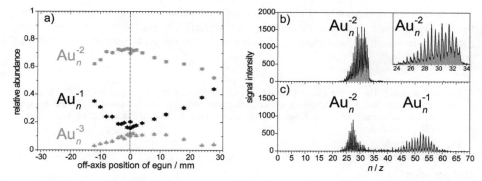

Fig. 6 **a** Relative yield of gold cluster polyanions (Au_n^{-z}, $n = 64 \pm 8$, $z = 1 - 3$) as a function of the radial filament position relative to the Penning trap axis. **b** ToF-spectrum of gold cluster dianions (Au_n^{-2}, $n = 48 - 68$) as produced in the RFQ II trap and subsequently stored in the Penning trap for 10 s. The distribution shows signals at full and at half-integer values of n/z due to the dianionic charge state $z = 2$ (inset). **c** ToF-spectrum of gold cluster anions after laser excitation of the dianions for 5 s (Nd:YAG laser, 532 nm, 30 Hz, 5 ns, 2.67 mJ). Monoanions are produced by photo-activated electron emission from the dianions

same side of the magnet, leaving the other side available for undisturbed cluster capture and time-of-flight analysis (Fig. 1) [17].

An example for laser excitation of gold cluster dianions is shown in Fig. 6. Here, the dianions were produced in the RFQ trap II prior to transfer into and storage in the Penning trap (Fig. 6b). After application of a pulsed Nd:YAG laser (repetition rate 30 Hz, wavelength 532 nm, pulse energy 2.67 mJ) for 5 s, monoanions are formed by electron emission from the dianions (Fig. 6c). Note that, in general, there is a size-dependent decay-pathway competition of electron emission and neutral-atom evaporation for dianions as well as further (possibly multiple) decay of the monoanions [37, 39]. In order to disentangle the details of the decay sequences, future experiments will include size and charge-state separation prior to the laser irradiation.

5 Summary and outlook

Different aspects of electron attachment to clusters in ion traps have been discussed with respect to polyanion production. Recent studies in a linear radio-frequency trap have been extended to the 3-state digital ion trap. This mode includes time slots with zero voltage between the RFQ rods, i.e. electrons encounter a field-free trapping volume. Thus, polyanion production with well-controlled low-energetic electrons will provide means for studying the repulsive Coulomb potential of negatively charged metal clusters.

While simultaneous storage of cluster monoanions and electrons in a Penning trap results in the formation of polyanionic clusters, the controlled removal of electrons from the trap will allow future lifetime studies of meta-stable polyanionic species. Moreover, an electron source located off the trap axis still provides an electron beam close to this axis due to the presence of the magnetic field of the Penning trap. Thus, the limited access to the trap volume inside the bore of the superconducting magnet remains available for application of, e.g., a laser beam.

25

Experimental results on gold clusters for polyanion production in both types of traps have been presented. In conclusion, ion traps are versatile and flexible tools for the investigation of size-selective metal clusters, in particular for studies on formation and stability of polyanions.

Acknowledgments The project was supported by the Collaborative Research Center of the DFG (SFB 652, TP A03). F. Martinez and S. Bandelow acknowledge postgraduate stipends from the state of Mecklenburg-Vorpommern (Landesgraduiertenförderung) in the framework of the International Max Planck Research School on Bounded Plasmas.

References

1. Scheller, M.K., Compton, R.N., Cederbaum, L.S.: Gas-phase multiply charged anions. Science **270**, 1160–1166 (1995)
2. Wang, X.-B., Wang, L.-S.: Experimental search for the smallest stable multiply charged anions in the gas phase. Phys. Rev. Lett. **83**, 3402–3405 (1999)
3. Yannouleas, C., Landman, U., Herlert, A., Schweikhard, L.: Multiply charged metal cluster anions. Phys. Rev. Lett. **86**, 2996–2999 (2001)
4. Schauer, S.N., Williams, P., Compton, R.N.: Production of small doubly charged negative carbon cluster ions by sputtering. Phys. Rev. Lett. **65**, 625–628 (1990)
5. Limbach, P.A., Schweikhard, L., Cowen, K.A., McDermott, M.T., Marshall, A.G., Coe, J.V.: Observation of the Doubly Charged Gas-Phase Fullerene Anions C_{60}^{2-} and C_{70}^{2-}. J. Am. Chem. Soc. **113**, 6795–6798 (1991)
6. Hettich, R.L., Compton, R.N., Ritchie, R.H.: Doubly charged negative ions of carbon C-60. Phys. Rev. Lett. **67**, 1242–1245 (1991)
7. Stoermer, C., Friedrich, J., Kappes, M.M.: Observation of Multiply Charged Cluster Anions upon Pulsed UV Laser Ablation of Metal Surfaces under High Vacuum. Int. J. Mass Spectrom. **206**, 63–78 (2001)
8. Hampe, O., Neumaier, M., Blom, M.N., Kappes, M.M.: On the generation and stability of isolated doubly negatively charged fullerenes. Chem. Phys. Lett. **354**, 303–309 (2002)
9. Liu, B., Hvelplund, P., Brøndsted Nielsen, S., Tomita, S.: Formation of C_{60}^{2-} dianions in collisions between C_{60}^{-} and Na atoms. Phys. Rev. Lett. **92**, 168301 (2004)
10. Hvelplund, P., Liu, B., Brøndsted Nielsen, S., Tomita, S.: Formation of higher-order fullerene dianions in collisions with Na atoms. Eur. Phys. J. D **43**, 133–136 (2007)
11. Hartig, J., Blom, M.N., Hampe, O., Kappes, M.M.: Electron attachment to negative fullerene ions: A fourier transform mass spectrometric study. Int. J. Mass Spectrom. **229**, 93–98 (2003)
12. Herlert, A., Krückeberg, S., Schweikhard, L., Vogel, M., Walther, C.: First observation of doubly charged negative gold cluster ions. Phys. Scr. T **80**, 200–202 (1999)
13. Schweikhard, L., Herlert, A., Krückeberg, S., Vogel, M., Walther, C.: Electronic effects in the production of small dianionic gold clusters by electron attachment on to stored Au_n^-, $n = 12 - 28$. Philos. Mag. B **79**, 1343–1352 (1999)
14. Walsh, N., Martinez, F., Marx, G., Schweikhard, L., Ziegler, F.: First observation of a tetra-anionic metal cluster, Al_n^{4-}. J. Chem. Phys. **132**, 014308–8 (2010)
15. Martinez, F., Bandelow, S., Breitenfeldt, C., Marx, G., Schweikhard, L., Wienholtz, F., Ziegler, F.: Appearance Size of Poly-Anionic Aluminum Clusters, Al_n^{z-}, $z = 2 - 5$. Eur. J. Phys. D **67**, 39–8 (2013)
16. Martinez, F., Bandelow, S., Marx, G., Schweikhard, L.: Production of multiply-charged metal-cluster anions in Penning and radio-frequency traps. AIP Conf. Proc. **1521**, 230–239 (2013)
17. Martinez, F., Bandelow, S., Breitenfeldt, C., Marx, G., Schweikhard, L., Vass, A., Wienholtz, F.: Upgrades at ClusterTrap and latest results. Int. J. Mass Spectrom. **365–366**, 266–274 (2014)
18. Martinez, F., Bandelow, S., Marx, G., Schweikhard, L., Vass, A.: Abundances of Tetra-, Penta- and Hexa-Anionic Gold Clusters. J. Phys. Chem. C (2015). doi:10.1021/jp510947p
19. Richards, J.A., Huey, R.M., Hiller, J.: On the time varying potential in the quadrupole mass spectrometer. Proc. IREE Aust. **32**, 321–322 (1971)
20. Richards, J.A., Huey, R.M., Hiller, J.: A new operating mode for the quadrupole mass filter. Int. J. Mass Spectrom. Ion Process. **12**, 317–339 (1973)
21. Ding, L., Sudakov, M., Kumashiro, S.: A simulation study of the digital ion trap mass spectrometer. Int. J. Mass Spectrom. **221**, 117–138 (2002)

22. Bandelow, S., Marx, G., Schweikhard, L.: The stability diagram of the digital ion trap. Int. J. Mass Spectrom. **336**, 47–52 (2013)
23. Bandelow, S., Marx, G., Schweikhard, L.: The 3-state digital ion trap. Int. J. Mass Spectrom. **353**, 49–53 (2013)
24. Martinez, F., Marx, G., Schweikhard, L., Vass, A., Ziegler, F.: The new ClusterTrap setup. Eur. Phys. J. D **63**, 255–262 (2011)
25. Schweikhard, L., Hansen, K., Herlert, A., Marx, G., Vogel, M.: New approaches to stored cluster ions. The determination of dissociation energies and recent studies on dianionic metal clusters. Eur. Phys. J. D **24**, 137–143 (2003)
26. Schweikhard, L., Krückeberg, S., Lützenkirchen, K., Walther, C.: The Mainz Cluster Trap. Eur. Phys. J. D **9**, 15–20 (1999)
27. Weidele, H., Frenzel, U., Leisner, T., Kreisle, D.: Production of "cold/hot" metal cluster ions: A modified laser vaporization source. Z. Phys. D **20**, 411–412 (1991)
28. Seidl, M., Perdew, J.P., Brajczewska, M., Fiolhais, C.: Ionization energy and electron affinity of a metal cluster in the stabilized jellium model: Size effect and charging limit. J. Chem. Phys. **108**, 8182–8189 (1998)
29. Herlert, A., Hansen, K., Schweikhard, L., Vogel, M.: Multiply charged titanium cluster anions: Production and photodetachment. Hyp. Int. **127**, 529–532 (2000)
30. Walsh, N., Martinez, F., Marx, G., Schweikhard, L., Ziegler, F.: Multiply negatively charged aluminium clusters II. Production of Al_n^{3-}. Eur. Phys. J. D **52**, 27–30 (2009)
31. Walsh, N., Martinez, F., Marx, G., Schweikhard, L.: Multiply negatively charged aluminium clusters. Production of Al_n^{2-} in a Penning trap. Eur. Phys. J. D **43**, 241–245 (2007)
32. Herlert, A., Jertz, R., Alonso Otamendi, J., Gonzalez Martinez, A.J., Schweikhard, L.: The influence of the trapping potential on the attachment of a second electron to stored metal cluster and fullerene anions. Int. J. Mass Spectrom. **218**, 217–225 (2002)
33. Martinez, F., Bandelow, S., Breitenfeldt, C., Marx, G., Schweikhard, L., Wienholtz, F., Ziegler, F.: Lifting of the trapping potential during ion storage for multi-anion production in a Penning trap. Int. J. Mass Spectrom. **313**, 30–35 (2012)
34. Walsh, N., Herlert, A., Martinez, F., Marx, G., Schweikhard, L.: Atomic clusters in a Penning trap: Investigation of their properties and utilization as diagnostic tools. J. Phys. B **42**, 154024-10 (2009)
35. Laude, D.A., Beu, S.C.: Suspended trapping pulse sequence for simplified mass calibration in Fourier transform mass spectrometry. Anal. Chem. **61**, 2422–2427 (1989)
36. Cooper, B.T., Buckner, S.W.: Simplified electron ejection in Fourier transform ion cyclotron resonance mass spectrometry by suspended trapping. Org. Mass Spectrom. **28**, 914–918 (1993)
37. Schweikhard, L., Hansen, K., Herlert, A., Herráiz Lablanca, M.D., Marx, G., Vogel, M.: Laser investigations of stored metal cluster ions. Hyp. Int. **146/147**, 275–281 (2003)
38. Herlert, A., Schweikhard, L.: First observation of delayed electron emission from dianionic metal clusters. Int. J. Mass Spectrom. **252**, 151–156 (2006)
39. Herlert, A., Schweikhard, L.: Two-electron emission after photoexcitation of metal-cluster dianions. New J. Phys. **14**, 055015-24 (2012)

27

Hyperfine Interact (2015) 236:29–37
DOI 10.1007/s10751-015-1196-y

Experimental study on dipole motion of an ion plasma confined in a linear Paul trap

K. Ito[1] · T. Okano[1] · K. Moriya[1] · K. Fukushima[1] ·
H. Higaki[1] · H. Okamoto[1]

Published online: 11 June 2015
© Springer International Publishing Switzerland 2015

Abstract The compact non-neutral plasma trap systems named "S-POD" have been developed at Hiroshima University as an experimental simulator of beam dynamics. S-POD is based either on a linear Paul trap or on a Penning trap and can approximately reproduce the collective motion of a relativistic charged-particle beam observed in the center-of-mass frame. We here employ the Paul trap system to investigate the behavior of an ion plasma near a dipole resonance. A simple method is proposed to calibrate the data of secular frequency measurements by using the dipole instability condition. We also show that the transverse density profile of an ion plasma in the trap can be estimated from the time evolution of ion losses caused by the resonance.

Keywords Linear Paul trap · Ion plasma · Dipole resonance · Beam dynamics

Pacs 29.20.-c · 41.75.-i · 52.27.Jt

1 Introduction

A novel application of a linear Paul trap has been proposed for systematic experimental studies of diverse beam dynamics issues [1–3]. The idea is based on the fact that a non-neutral plasma confined in this type of trap is physically almost equivalent to a charged particle beam propagating through a modern accelerator. Employing this idea, we designed and constructed the compact experimental facility called S-POD (Simulator of Particle Orbit

Proceedings of the 6th International Conference on Trapped Charged Particles and Fundamental Physics (TCP 2014), Takamatsu, Japan,1-5 December 2014

✉ K. Ito
   kzito@hiroshima-u.ac.jp

1   Graduate School of Advanced Sciences of Matter, Hiroshima University, 1-3-1 Kagamiyma, Higashi-Hiroshima, Hiroshima 739-8530, Japan

Dynamics) at Hiroshima University [4–6]. There are many practical advantages in S-POD experiment [1]. For instance, the system is much more compact, flexible, and cheaper than large-scale accelerators and beam transport channels. It is easier to observe the behavior of a plasma in a trap because its centroid is at rest in the laboratory frame. Unlike accelerator-based experiments, we do not have to worry about radio-activation due to particle losses. Princeton Plasma Physics Laboratory also developed a similar trap system for beam physics applications [7]. Their system is also a Paul trap but employs three colinear cylinders, each of which is divided into four 90° sectors to provide a quadrupole plasma confinement field. The dimension is much lager than ours. The central cylinder is 2 m long and 0.2 m in diameter.

In general, charged-particle beams are focused by a series of quadrupole focusing and defocusing magnets periodically aligned along the design beam orbit. As is well-known, the periodic nature of the external focusing potential excites resonant beam instability under a certain condition. Since any beam focusing channels inevitably contain error fields, various extra linear and nonlinear resonances can occur in real machines. In particular, the integer resonance driven by a dipole error field is of practical importance because it is the lowest order and thus strongly affects the beam stability. The unique feature of this type of resonance is that the instability condition is independent of beam density; namely, particle losses occur at any beam density when the *betatron tune*, which corresponds to the secular frequency, is close to an integer. A regular Paul trap is, however, free from integer resonance due to the lack of the dipole driving field except for specific cases [8].

In a recent experimental study with S-POD [9], we intentionally introduced periodic perturbing voltages of dipole symmetry to the quadrupole electrodes of the linear Paul trap. We then confirmed the excitation of integer resonance and localized ion losses independent of plasma density. The original purpose of that study was to deepen the understanding of integer resonance crossing in a non-scaling fixed-field alternating gradient accelerator [10]. The present paper addresses a simple technique to figure out the effective plasma confinement strength, in other words, the secular frequency or the *bare* betatron tune, by using measured ion-loss data. This parameter is most important in beam dynamics, but in actual Paul-trap experiments, the accuracy of tune determination depends on the performance of measurement devices (e.g. an oscilloscope) and machining accuracy of the trap. Since it is guaranteed that the dipole instability occurs exactly on an integer tune regardless of the plasma density, we can calibrate the linear focusing strength and accurately determine the bare tune. We also show that the time-evolution of ion losses on an integer resonance can be employed to deduce the transverse spatial profile of an ion plasma before the resonance is excited.

2 Betatron tune and resonance condition

A large circular accelerator, such as a synchrotron or a storage ring, is generally composed of several identical focusing structures called *superperiods*. Each superperiod often contains more than one alternating focusing (AG) blocks that we call here a *unit cell*. The most standard cell structure is the so-called "FODO" that includes a single focusing quadrupole and a single defocusing quadrupole. In a regular Paul trap that uses the sinusoidal electric field at the frequency of f_Q to confine ions, a single RF cycle corresponds to a single FODO period. The transverse x motion of a single particle in either system obeys the Hill's equation whose general solution can be written, according to Floquet theorem, as $x = \text{const} \times \sqrt{\beta} \cos \psi$ where β is the so-called betatron function satisfying

$$\frac{d^2 \sqrt{\beta}}{d\tau^2} + K_Q(\tau)\sqrt{\beta} - \frac{1}{(\sqrt{\beta})^3} = 0, \tag{1}$$

where $\tau = ct$ with c being the speed of light and $K_Q(\tau)$ is proportional to RF voltages applied to the quadrupole electrodes. The phase function ψ can be related to β as $d\psi/d\tau = 1/\beta$. Assuming a circular machine composed of N_{cell} identical FODO blocks, the betatron tune around the ring is calculated from [11]

$$\nu_0 = \int_0^{N_{cell}\tau_0} \frac{d\tau}{\beta(\tau)} = N_{cell} \int_0^{\tau_0} \frac{d\tau}{\beta(\tau)} = N_{cell} \frac{f_s}{f_Q}, \tag{2}$$

where $\tau_0 = c/f_Q$, and f_s is the frequency of the secular motion. The RF waveform of $K_Q(\tau)$ directly reflects the AG focusing structure of a particular accelerator. The RF power control system of S-POD can generate a variety of periodic waveforms to explore beam dynamic effects in many different AG lattice configurations. Considering the simple sinusoidal waveform with the voltage amplitude V_Q, we have

$$K_Q = \frac{2q_i V_Q}{M_i c^2 r_0^2} \cos\left(\frac{2\pi f_Q}{c}\tau\right), \tag{3}$$

where M_i and q_i are the mass and charge state of confined ions, and r_0 is the radius of the trap aperture.

According to a one-dimensional Vlasov theory [12], the resonant instability of order m is expected to take place under the condition $m(\nu_0 - C_m \Delta\nu) \approx nN_{sp}/2$, where $\Delta\nu$ is the tune shift caused by the Coulomb repulsive force, C_m is a m-dependent constant less than unity, n is an integer corresponding to the Fourier harmonic number of the external periodic driving force, and N_{sp} is the number of superperiods around the ring. This condition applies to the instabilities of the quadrupole mode ($m = 2$) or higher-order nonlinear modes ($m > 2$). As already mentioned, the dipole-mode instability ($m = 1$) is independent of the Coulomb potential; the resonance condition is simply $\nu_0 \approx nN_{sp}$. Note also that the factor $1/2$ is missing on the right hand side because of the peculiarity of the dipole resonance.

3 Experimental setup

The structure of a linear Paul trap for S-POD is illustrated in Fig. 1 [4, 5]. The trap is axially divided into five quadrupole sections, namely, End A, IS, Gate, ER, End B, that are electrically isolated from each other so that we can add different DC bias voltages. The lengths of the End and Gate sections are 14 mm, while the IS and ER sections are 75 mm long. The radius of the plasma confinement region is $r_0 = 5$ mm and the radius of quadrupole rods is $r_e = 1.15 r_0 = 5.75$ mm. The trap is placed in a vacuum chamber with the base pressure of 8×10^{-8} Pa.

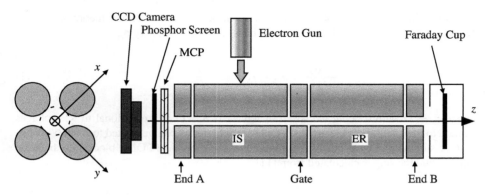

Fig. 1 Schematic view of the multisection linear Paul trap

To generate $^{40}Ar^+$ ions, neutral Ar gas ($< 2 \times 10^{-5}$ Pa) is introduced into the vacuum chamber and ionized with an electron beam (135 eV, < 0.5 mA) in the IS region for 0.2 s. Those ions are confined radially by a quadrupole RF field ($V_Q < 94$ V, $f_Q = 1$ MHz) and axially by DC bias voltages (30 V) on the End and Gate electrodes. After a sufficient number of ions are stored, perturbing RF voltages V_D with opposite signs are added to a pair of rods in the IS region to excite the dipole oscillation in the x-direction for a time period of t_D. The DC bias on either the End or Gate section is dropped to extract the plasma from the IS section. Extracted ions eventually reach either a Faraday cup or a micro-channel plate (MCP) with a phosphor screen where we can measure the total number N_i of ions surviving after t_D. If the plasma is unstable, N_i will be reduced significantly. The Faraday cup is used for a plasma of relatively high density ($N_i \geq 10^5$) because of the limited S/N ratio. In the range $N_i < 10^5$, we employ the MCP that enables us to obtain the transverse plasma profile as well with a CCD camera [5]. The longitudinal plasma profile should approximately be uniform, because the longitudinal DC potential well is almost flat over the IS region. Need-less to say, the plasma as a whole executes a quadrupole oscillation on the $x - y$ plane due to the quadrupole focusing potential. Once the dipole driving field is switched on, a rigid dipole oscillation is superimposed on the quadrupole motion. The frequencies of both oscil-lations are much shorter than the time period within which all extracted ions are damped at the MCP (> 100 μs). This means that we can only measure the average transverse profile integrated along the axial direction over a long extraction period. Note also that the MCP is placed about 10 mm away from the trap end. Since the transverse focusing force disappears in this end region, the plasma is radially expanded to some degree. We thus need to estimate the transverse expansion factor in order to figure out the plasma profile before extraction. A three-dimensional particle tracking code has been developed for this purpose.

4 Tune calibration and transverse profile measurements

Arbitrary numbers can be chosen for N_{cell} and N_{sp} in S-POD experiments. Among a wide range of choices, we here assume $N_{cell} = 42$ and $N_{sp} = 1$ recalling our most recent work on dipole resonance [9]. These numbers reflect the situation of the EMMA accelerator con-sisting of 42 AG focusing cells [10]. Ideally, the ring has 42-fold symmetric structure, but a localized dipole leakage field from septum magnets destroys the high lattice symmetry reducing the superperiodic number down to 1 [13]. Figure 2a schematically represents the

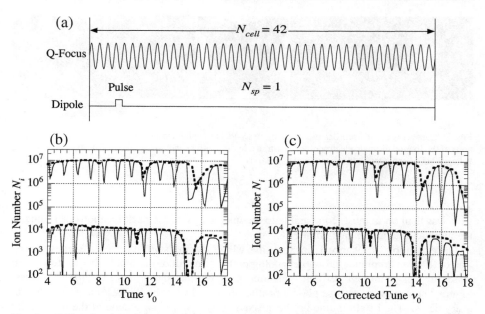

Fig. 2 **a** Schematic of the focusing and dipole RF waveforms corresponding to the EMMA situation. **b** Resonance stop bands observed with (*solid*) and without (*dotted*) the pulsed dipole perturbation. **c** The stop-band distribution after the tune calibration procedure

RF waveforms employed for the present S-POD experiment. A dipole pulse, 1 μs in width and 1 V in height, is applied every 42 quadrupole focusing periods to excite integer resonances. The solid and dotted lines in Fig. 2b show the stop-band distributions measured, respectively, with and without the dipole perturbation. The number of ions surviving after $t_D = 10$ ms is plotted as a function of bare tune ν_0. The tune can be varied over a wide range simply by changing the amplitude of the RF voltages of the quadrupole rods. Two different initial ion numbers are considered in this example. The serious ion losses observed near $\nu_0 = 42/4$ and 42/3 even without the dipole driving force are caused mainly by the resonant instabilities of the quadrupole ($m = 2$) and sextupole ($m = 3$) modes, respectively. Note that these stop bands shift rightward with higher N_i because $\Delta\nu$ becomes larger. By contrast, many additional stop bands induced by the dipole perturbation do not move depending on the initial number of ions. We, however, recognize that these stop bands of dipole resonance are slightly deviated from integer tunes calculated from (1)–(3) with raw data. In order to evaluate the bare tune at each operating point, we first save the actual RF waveform measured by an oscilloscope. We then substitute the measured data in (1) to obtain the stationary solution numerically and use (2) for tune determination. This should be due partly to possible systematic errors included in RF waveform measurements with an oscilloscope. A slight discrepancy between the design r_0 and actual r_0 due to a mechanical error is another source of dipole stop-band shifts. These errors can be corrected largely by just multiplying the focusing function $K_Q(\tau)$ by a constant factor α. Making use of the fact that the dipole resonance always occurs at an integer tune, we can readily determine the correction factor by a least-squares fit. In the present case, we conclude that α with a standard error is 0.958 ± 0.001. The data in Fig. 2b is replotted in Fig. 2c with this correction factor. All dipole stop bands are now located at integer tunes. The root-mean-squared residual error in

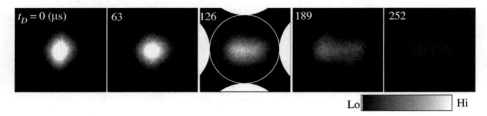

Fig. 3 Transverse plasma profiles measured by the S-POD imaging system at different timings. The positions of the four quadrupole electrodes are illustrated in the central panel for reference. The dipole perturbing voltages have been applied to the two horizontal rods

ν_0 is less than or, at most, comparable to the uncertainty of the peak locations of ν_0 due to the interval of adjacent data points.

Once a dipole resonance is excited, the whole plasma starts to oscillate transversely. The oscillation amplitude grows depending on the strength of the dipole driving force. Ion losses are caused through periodic collisions of the plasma edge with the electrode surfaces. Figure 3 shows the transverse plasma profiles observed with the S-POD imaging system at every 63 μs. The betatron tune has been fixed at $\nu_0 = 8$. The amplitude of the 8th Fourier harmonic of the dipole perturbation driving this resonance is $V_D = 0.1$ V. We started this experiment at a relatively low number of ions ($N_i = 3.2 \times 10^5$) to minimize complex collective effects. The integrated image of the plasma profile is gradually expanded in the horizontal direction because of the rapid dipole oscillation and eventually fades away.

Since the speed of ion losses on an integer resonance depends not only on the dipole perturbation strength but also on the transverse plasma extent, we can deduce the average radius of the initial ion distribution from the time evolution of ion losses. In Fig. 4a, the fraction of surviving ions S evaluated from simple one-dimensional (1D) tracking simulations is plotted with the dashed curve for comparison with experimental data. In the 1D simulations, a Gaussian distribution of ions with the temperature of 0.5 eV has been assumed on the basis of past S-POD experiments [5]. We see that the experimental observation is in very good agreement with the numerical prediction.

Ideally, the tune of the plasma centroid oscillation on integer resonance coincides with that of the single-particle betatron motion. This suggests that roughly a half of initially confined ions are scraped by the horizontal electrodes in Fig. 3 when the transverse shift of the plasma centroid reaches the aperture radius r_0. We have verified this expectation through numerical simulations. The amplitude of the centroid oscillation, x_g, can be estimated from a driven harmonic oscillator model that gives $x_g = u_D \zeta$, where u_D is the scaled linear growth rate and $\zeta = V_D t_D$, the abscissa of Fig. 4a. The use of x_g allows us to approximate the loss function S from

$$S(x_g) = \frac{L_p}{N_0} \int_{-r_0}^{r_0} dy \int_{-r_0}^{r_0 - x_g} \rho(x, y) \, dx = \int_{-r_0}^{r_0 - x_g} \xi(x) \, dx, \qquad (4)$$

where N_0 and ρ are the ion number and the ion density distribution when $x_g = \zeta = 0$, L_p is the effective length of the plasma, and ξ is the normalized plasma profile projected to the x-axis. Considering that the time evolution of the measured S in Fig. 4a is similar to the

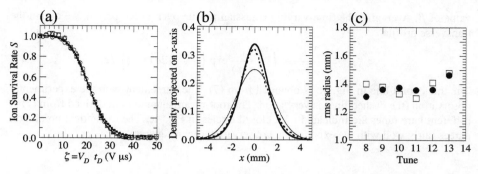

Fig. 4 a Fraction of $^{40}Ar^+$ ions surviving on the integer resonance at $\nu_0 = 8$. The abscissa is the scaled plasma-confinement time defined as $V_D \times t_D$. Open *Circles*, *squares*, and *triangles* represent measurement data obtained, respectively, with $V_D = 0.5$, 1, and 2 (V). The *solid line* is obtained from the fitting formula in (6). The *dashed line* almost overlapping with the *solid curve* is obtained from a 1D particle tracking simulation. **b** Normalized ion profiles projected on x - axis. The *thin solid curve* is the Gaussian distribution defined by (5). The *thick solid curve* represents the average profile given by the formula in (7). The *dashed line* shows the experimental observation by the imaging system. **c** Transverse rms extent of an ion plasma at different tunes. Closed *circles* are the theoretical predictions based on (7) while open *squares* represent experimental data

error function, we here adopt the following Gaussian profile ξ_f within the range $|x| < r_0$ as the initial ion distribution:

$$\xi_f(x) = \frac{a}{\sqrt{2\pi}\sigma_0} \exp\left[-x^2/(2\sigma_0^2)\right],$$ (5)

where $a = 1/\text{erf}\left[r_0/(\sqrt{2}\sigma_0)\right]$ is the normalization constant to fulfill $\int_{-r_0}^{r_0} \xi_f dx = 1$. The fitting function for the surviving ion fraction can then be expressed as

$$S_f(x_g) = \frac{1}{2}\left\{1 + a\,\text{erf}\left[(r_0 - x_g)/(\sqrt{2}\sigma_0)\right]\right\}.$$ (6)

The solid line plotted in Fig. 4a is the fitting curve based on this formula with $u_D = 0.245$ mm/(Vμs) and $\sigma_0 = 1.69$ mm. We also tried a high-order power series as a fitting function for ξ_f, but the result was almost identical to the Gaussian prediction.

The Gaussian profile ξ_f is plotted in Fig. 4b with the thin solid line while the dashed line shows the ion distribution experimentally observed with the S-POD imaging system when $x_g = \zeta = 0$. We find that the simple fitting model based on the measured ion-loss curve in Fig. 4a does not well explain the actual profile on the MCP. The discrepancy probably comes from the fact that the transverse distribution is not static but constantly driven by the external RF quadrupole potential. The observed profile on the MCP has been averaged over many periods of the quadrupole oscillation, which distorts the distribution from the simple Gaussian.

In order to improve the theoretical prediction, we incorporate the transverse quadrupole motion of the plasma into the present Gaussian model. At the low ion density considered here, the collective Coulomb potential is negligible. Then, the approximate plasma extent can be expressed by $\sqrt{\epsilon\beta(\tau)}$ where ϵ is a constant corresponding to the plasma emittance and $\beta(\tau)$ is the periodic solution of (1). The effect of the quadrupole oscillation can be taken into account by replacing σ_0 in (5) by $\sigma(\tau) = \sigma_0\sqrt{\beta(\tau)/\beta_0}$, where β_0 is the maximum value of β. Averaging the time-varying Gaussian profile over an RF period, we obtain the transverse profile

$$< \xi_f > = \frac{a}{\sqrt{2\pi}\tau_0} \int_0^{\tau_0} \frac{1}{\sigma(\tau)} \exp\left\{-x^2/[2\sigma(\tau)^2]\right\} d\tau. \tag{7}$$

The thick solid line in Fig. 4b is obtained from (7). The agreement with the experimental observation (the dashed line) is very good. The root-mean-squared radii derived from (7) at different bare tunes are indicated with closed circles in Fig. 4c. The theoretical prediction agrees fairly well with the experimental observation (open squares)

5 Summary

We successfully excited integer resonance stop bands, applying a sinusoidal or a pulse dipole perturbation to the quadrupole electrodes of a linear Paul trap. Sharp ion losses were observed near integer tunes corresponding to the frequencies of the driving Fourier harmonics. We confirmed that the stop bands of the dipole-mode instability do not move depending on the plasma density. Slight shifts of the stop-band locations from integer tunes were, however, found which are most likely caused by technical limitations such as misalignments of the electrodes, possible errors in RF voltage measurements, etc. Taking advantage of the integer resonance condition, we evaluated a calibration factor to determine the betatron tune (secular frequency) precisely.

The time evolution of ion losses on an integer resonance was also measured and compared with numerical results from a simple particle-tracking simulations. We showed that it is possible to make a good estimate of the transverse plasma profile by using the ion-loss data on resonance.

Acknowledgments This work was supported by the JPSJ KAKENHI Grant No. 24340053.

Conflict of interests The authors declare that they have no conflict of interest.

References

1. Okamoto, H., Tanaka, H.: Proposed experiments for the study of beam halo formation. Nucl. Instrum. Methods Phys. Res. A **437**, 178 (1999)
2. Davidson, R.C., Qin, H., Shvets, G.: A Paul trap configuration to simulate intense non-neutral beam propagation over large distances through a periodic focusing quadrupole magnetic field. Phys. Plasmas **7**, 1020 (2000)
3. Okamoto, H., Wada, Y., Takai, R.: Radio-frequency quadrupole trap as a tool for experimental beam physics. Nucl. Instrum. Methods Phys. Res. A **485**, 244 (2002)
4. Takai, R., Enokizono, H., Ito, K., Mizuno, Y., Okabe, K., Okamoto, H.: Development of a compact plasma trap for experimental beam physics. Jpn. J. Appl. Phys. **45**, 5332 (2006)
5. Ito, K., Nakayama, K., et al.: Determination of transverse distributions of ion plasmas confined in a linear Paul trap by imaging diagnostics. Jpn. J. Appl. Phys. **47**, 8017 (2008)
6. Ohtsubo, S., Fujioka, M., Higaki, H., Ito, K., Okamoto, H., et al.: Experimental study of coherent betatron resonances with a Paul trap. Phys. Rev. STAB **13**, 044201 (2010)
7. Gilson, E.P., Davidson, R.C., Efthimion, P.C., et al.: Excitation of transverse dipole and quadrupole modes in a pure ion plasma in a linear Paul trap to study collective processes in intense beamsa. Phys Plasma **20**, 055706 (2013)
8. Schwartz, J.C., Senko, M.W., Syka, J.E.: A two-dimensional quadrupole ion trap mass spectrometer. J. Am. Mass Spectrom **13**, 659 (2002)

9. Moriya, K., Fukushima, K., Ito, K., Okano, T., Okamoto, H., et al.: Experimental study of integer resonance crossing in a non-scaling fixed field alternating gradient accelerator with a Paul ion trap. Phys. Rev. STAB **18**, 034001 (2015)
10. Machida, S., Barlow, R., Berg, J.S., et al.: Acceleration in the linear non-scaling fixed-field alternating-gradient accelerator EMMA. Nat. Phys. **8**, 243 (2012)
11. Lee, S.Y.: Accelerator physics, pp. 42–73. World Scientific (2004)
12. Okamotoa, H., Yokoya, K.: Parametric resonances in intense one-dimensional beams propagating through a periodic focusing channel. Nucl. Instrum. Methods Phys. Res. A **482**, 51 (2002)
13. Sheehy, S.L., Kelliher, D.J., et al.: Experimental studies of resonance crossing in linear non-scaling FFAGs with the S-POD Plasma trap. Proc. IPAC2013, 4056 (2013)

Hyperfine Interact (2015) 236:39–51
DOI 10.1007/s10751-015-1209-x

Development of multiple laser frequency control system for Ca$^+$ isotope ion cooling

**Kyunghun Jung[1] · Yuta Yamamoto[2] ·
Shuichi Hasegawa[1]**

Published online: 29 December 2015

Abstract We here developed and evaluated a laser frequency control system which synchronizes the laser frequency to the resonance of target Ca$^+$ isotope ion whose having more than 8 GHz of isotope shift based on the Fringe Offset Lock method for simple operation of ICPMS-ILECS (Inductively Coupled Plasma Mass Spectrometry - Ion trap Laser Cooling Spectroscopy) The system fulfilled the minimum requirements of four slave lasers stability for Doppler cooling of Ca$^+$ ions. A performance of the system was evaluated by cooling ^{40}Ca$^+$ ions with the stabilized slave lasers. All the stable even Ca$^+$ isotope ions were trapped and their fluorescence was observed by switching laser frequencies using the system. An odd calcium isotope ^{43}Ca$^+$ cooling was also succeeded by the control system.

Keywords Ion trap · Calcium ion · Isotope · Diode laser · Fringe offset lock

Proceedings of the 6th International Conference on Trapped Charged Particles and Fundamental Physics (TCP 2014), Takamatsu, Japan, 1–5 December 2014

✉ Kyunghun Jung
jung@lyman.q.t.u-tokyo.ac.jp

Yuta Yamamoto
yamamoto@lyman.q.t.u-tokyo.ac.jp

Shuichi Hasegawa
hasegawa@tokai.t.u-tokyo.ac.jp

[1] Nuclear Professional School, The University of Tokyo, 2-22 Shirane, Shirakata, Tokai, Naka, Ibaraki 319-1188, Japan

[2] Department of Nuclear Engineering and Management, The University of Tokyo, 7-3-1 Hongo, Bunkyo-ku, Tokyo 113-8656, Japan

1 Introduction

An easy downsizing and low cost operation are available for semiconductor lasers because it requires only a small current for its driving and can be thermally stabilized by a small Peltier element [1]. For these reasons ring and dye laser have been gradually replaced with semiconductor lasers in atomic spectroscopy. Nowadays semiconductor lasers are widely used in a field of absorption, fluorescence and ionization spectroscopy [2]. In the atomic spectroscopy, a laser, single mode linewidth of which is narrower than the natural linewidth of target atoms is preferred to resolve its spectra. Therefore various techniques have been developed for narrowing linewidths of semiconductor lasers [3, 4]. A wellknown method is to form an external cavity outside a laser diode [5]. This External Cavity Diode Laser (ECDL) [6] can easily obtain a stabilized single mode because of its simple structure. For a high precision spectroscopy a laser frequency stabilization is very important because it fluctuates due to unstable driving current temperature variation, mechanical vibrations and so on. There are various methods to stabilize a laser frequency, such as, using an atomic absorption line as an absolute frequency reference [7–10], judging a relative frequency of a laser by an etalon [11, 12], detecting a phase difference between a reference and a control laser by mixing [13, 14], transferring a stability of reference laser to the other lasers through an etalon [15, 16].

As considering requirements of laser cooling of ions, the stabilization method using an atomic absorption line requires the line in the vicinity of the cooling ion resonance frequency and Acousto-Optic Modulators (AOM) are also needed to sweep the laser frequency. In case of using the phase difference stabilization, the method cannot be achieved without an expensive modulation equipment for each individual laser. However the Fringe Offset Lock (FOL) method is able to control multiple lasers using only one etalon and frequency sweep of the slave lasers can be easily done. Therefore the method can be think as very useful for the ion trap experiment. The FOL method was proposed by Lindsay [17]. A stability of the method was increased by controlling a temperature of a cavity [18] or housing an etalon in a sealed chamber [19] or increasing a feedback bandwidth with a high speed scanning of an etalon [20]. Also the method was advanced by solving a control stop problem when fringe signal was moved more than one FSR of etalon [21] and realizing multiple laser controlling [22].

In the ICPMS-ILECS (Inductively Coupled Plasma Mass Spectrometry - Ion trap Laser Cooling Spectroscopy) [23], cooling lasers frequency has to be synchronized with the resonance frequency of target ions which were injected from the ICPMS to trap and cool. Although an isotope of ion beams can be easily selected by the ICPMS, manual switching and stabilization of the cooling lasers frequency corresponding to the isotope shift of selected target ions is a troublesome task. Therefore we developed and evaluated a DFOC (Digital Fringe Offset Control) system based on the FOL method [24] to synchronize the laser frequency on the resonance of the target ion. Two 397 nm and two 866 nm slave ECDL frequencies can be simultaneously stabilized and tuned on the resonance of target isotope ion by the system. The DFOC system stabilizes the slave lasers frequency by chasing its fringe positions continuously and only one etalon is needed to control and monitor frequency of all the slave lasers And the system does not requires a lock-in amplifier to create an error signal therefore an entire laser cooling system can be downsized. Because the system is able to chase the fringe which moved more than a Free Spectral Range (FSR) of the etalon a frequency control range of the system will not be limited by the FSR. A details of the DFOC system construction and its performance will be explained in this paper.

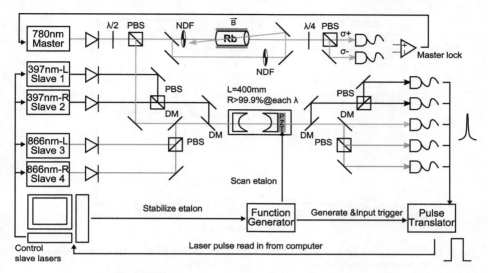

Fig. 1 Optical setup and signal flows for the DFOC system. The master laser beam was divided for the SD-DAVLL and obtaining a fringe signal. All the laser beams were spatially overlapped and separated using their frequencies and polarizations. Fringe signals were converted into TTL pulses and a computer created a feedback signal. (DM : Dichroic mirror, PBS : Polarized beam splitter, NDF : Neutral density filter, λ/2 and λ/4 : Waveplate)

2 Experimental setup

2.1 System requirement for Doppler cooling of ion and its scheme

DFOC is a multiple laser frequency control system which is used for Doppler cooling of trapped Ca^+ ions in this experiment $4s\,^2S_{1/2} \leftrightarrow 4p\,^2P_{1/2}$ or $4p\,^2P_{3/2}$ transitions can be used for Doppler cooling of Ca^+ ions and in our case $4s\,^2S_{1/2} \leftrightarrow 4p\,^2P_{1/2}$ transition was chosen since the transition requires only 397 and 866 nm lasers to close the cooling cycle though the latter case requires an additional 854 nm laser A frequency of cooling laser is needed to be red detuned by a half amount of the natural linewidth Γ of the cooling transition for an efficient cooling [25]. Because the Γ of $4s\,^2S_{1/2} \leftrightarrow 4p\,^2P_{1/2}$ transition is 22.4 MHz [26], the 397 nm laser frequency has to be 11.2 MHz red detuned from the resonance of Ca^+ ions. If the cooling laser frequency is larger than the resonance of a target ion, the ion is heated and a cooling efficiency will be decreased. Therefore a frequency fluctuation of the 397 nm laser have to be at least smaller than 22.4 MHz to not to heat ions. An 866 nm laser is needed to repump the $3d\,^2D_{3/2} \leftrightarrow 4p\,^2P_{1/2}$ transition. Because a lorentzian component of the transition linewidth is experimentally measured as 30 MHz [27], the frequency fluctuation of 866 nm laser has to be less than 30 MHz.

The FOL method obtains fringe signals of the master and the slave lasers from a same etalon and stabilizes the slave laser frequency by maintain a width between the two laser fringes. The DFOC transforms the fringe signals into TTL pulses to suppress an electrical noise and handle the signals easily by a computer. An optical setup and a signal flow of the control system is shown in Fig. 1. All the laser beams were spatially overlapped and then

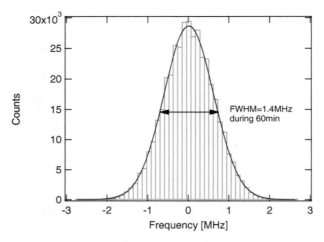

Fig. 2 The 780 nm laser stabilization result by the SD-DAVLL method for 60 minutes. The error signal was converted into a frequency and its histogram was created from the data

separated using their wavelength and polarizations in the system because the fringe signals of all the lasers have to be obtained individually. Fringes of the five lasers from the etalon is transformed into the pulse signal and acquired by the computer. The computer creates feedbacks based on the pulse signals to control the etalon and the slave lasers.

2.2 Master laser

Since a stability of the master laser is transferred to those of the slave lasers in the FOL method, usually a commercial He-Ne laser or a stabilized semiconductor laser is used as a master laser to obtain high stability An ECDL which is stabilized by Sub-Doppler Dichroic Atomic Vapor Laser Lock (SD-DAVLL) [28] was introduced as the master laser in the DFOC system because a cheap laser diode could be used and better stability than the stabilized He-Ne could be obtained. The reason why the SD-DAVLL was used is the technique hardly affected from a temperature variation of an atomic cell [29] and no laser modulation is required to obtain an error signal. The SD-DAVLL uses the shifting of absorption frequencies by Zeeman effect when a saturated absorption spectroscopy is conducted with an atomic cell in magnetic fields. The shifted components can be distinguished by dividing a probe beam into right and left-circularly polarized components and the error signal for frequency stabilization is obtained from a difference of the two circularly polarized component signals. The upper part of Fig. 1 shows an optical setup and a feedback flow for the SD-DAVLL The wavelength of the master laser was chosen as a 780 nm to use the Rubidium atom D2 line and the master laser frequency was stabilized with the SD-DAVLL using the ^{85}Rb absorption line. An error signal of the SD-DAVLL was recorded for an hour and converted into a histogram The result was shown in Fig. 2. A stability of the frequency locked master laser was measured as 1.4 MHz of Full Width at Half Maximum (FWHM) during one hour. The stability of our master laser is comparable to those of typical commercial frequency stabilized He-Ne lasers.

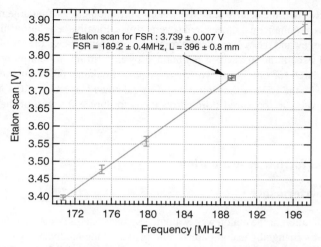

Fig. 3 A calibration curve for the etalon FSR measurement. Its x-axis is a modulation amount corresponds to the measured beat frequency and a y-axis does to a ring piezo scan voltage. The FSR of etalon was measured as 189.2 ± 0.4 MHz from the curve

2.3 Measurement of the etalon FSR

The DFOC system estimates a relative frequency of a laser by converting its fringe signal movement into a frequency scale using the Free Spectral Range (FSR) of the etalon. The DFOC system introduced a confocal type lab made etalon whose curvature radius r of its mirror is 400 mm therefore the FSR can be calculated as 187.4 MHz. The mirror used our etalon was produced to have more than 98 % of reflectance for 397, 780 and 866 nm. A finesse of fringe signal was adjusted to be around 10 to 30 to obtain a better stability of the laser. However we experimentally measured the FSR using a laser because a correct value of the etalon FSR is required for precise laser frequency control For the measurement, a fundamental 866 nm beam and its modulated beam were prepared. The fundamental beam was divided by a beam splitter and one of the divided beams was modulated by an AOM. The fundamental and modulated beams were overlapped spatially then divided into the two beams. The half of the overlapped beam was guided into an avalanche photodiode (Hamamatsu photonics, C5658) to measure its beat frequency and the other one to the etalon. The beat frequency from the avalanche photodiode was measured by a spectrum analyzer (Advantest, U3751). Because the two fundamental and modulated beams were originated from the same 866 nm laser, the frequency jitter of the laser was balanced out and an absolute amount of frequency modulation by the AOM could be measured with a precision of less than 10 kHz FWHM. At the same time the fringe peak position difference of fundamental and modulated beam was measured as a function of an applied voltage on a ring piezo element for the cavity scan. Therefore a relationship between the cavity length and an absolute frequency could be measured from the result and finally the FSR could be calculated.

The measured data and the calibration curve is shown in Fig. 3. First, the modulation amount was adjusted to be $170 \sim 200$ MHz which was in the vicinity of designed value of the etalon FSR. Then the scan voltage between the fringe peak of fundamental and modulated beam was simultaneously measured several times. Finally the relationship between the absolute frequency and the voltage on the ring piezo element to scan the absolute frequency were obtained A calibration curve was obtained by line fitting of data points Because the etalon scan voltage for obtaining two fringe peak of fundamental beam was measured as 3.739 ± 0.007 V, the FSR of the etalon could be calculated as 189.2 ± 0.4 MHz by referring to the curve. The measured FSR corresponds to 396 ± 0.8 mm of the actual cavity length.

3 System performance evaluation

3.1 System performance

At first a stabilization of the etalon was evaluated by the system. Because if the cavity length of the etalon was changed, the computer would give a wrong feedback signal to slave lasers in the DFOC system. Therefore the system compensates a changed cavity length due to an environment condition by stabilizing a fringe peak position of the master laser. Our master laser frequency measured by etalon was drifted a few hundred of MHz according to the experiment room temperature fluctuation although the laser frequency was stabilized by SD-DAVLL. Around 2 °C of the temperature variation was occurred within less than 10 minutes due to the self-adjustment of an air conditioner in the experiment room. However the fringe position of the master laser was stabilized in few MHz scale although the room temperature varied greatly during the etalon length compensation by the system. From these result we can conclude that the system was succeeded to compensate the cavity length variation for increasing a stability of the system.

Then the slave laser frequency stabilization performance was evaluated. The DFOC system is able to control a frequency of two 397 nm and two 866 nm slave lasers simultaneously. Few hundred MHz of the laser frequency drift was appeared during 30 minutes in our setup. Then the stabilized fringe positions of four slave lasers were measured during 50 minutes. A stabilized laser frequency linewidth was measured as 6.2 and 3.5 MHz for 397 nm-L and R lasers And for 866 nm-L and R lasers the stability was measured as 6.8 and 4.3 MHz. The stability difference of lasers can be thought of as a result from a various mechanical precision of lab-made lasers and different drivers for laser to laser, therefore the noise scale of lasers are all different.

And we again evaluated a frequency stability of the 866 nm slave lasers using an optical heterodyne method to find the stabilization result obtained by the etalon was precise or not. For the measurement, the two 866 nm slave lasers were stabilized by the system and its beat frequency was measured around one hour. More than 30 MHz of the frequency drift was appeared on the measurement result within less than 15 minutes. The FWHM of the measured beat frequency during the stabilization period of only 67 minutes was 5.3 MHz From these result we can say the frequency evaluated by the etalon having enough precision for experiment because a precision needed for ion Doppler cooling is only ten MHz scale. Consequently we can conclude that the enough stability for Ca$^+$ ion Doppler cooling was obtained for the four slave lasers by the simultaneous frequency stabilization control of the DFOC system.

Table 1 Natural abundance of Calcium isotopes and their isotope shifts [31]

| Mass number (Z) | Natural abundance (%) | Isotope shift (MHz) | |
		Ca$^+$ S-P 397 nm	Ca$^+$ D-P 866 nm
40	96.9	0	0
42	0.647	425	−2350
43	0.135	688	−3465
44	2.09	842	−4495
46	0.004	1287	−6478
48	0.187	1696	−8288

3.2 Ion trap experiment

To estimate the Doppler cooling performance of the constructed DFOC system ^{40}Ca$^+$ ion was selected as a target ion. The ions were trapped and cooled in the ICPMS-ILECS which is a mass spectrometry using a linear Paul trap technique. A commercial ICPMS was adopted here as an ion source because of its high ionization efficiency for various elements and a continuous liquid sample injection ability from the atmosphere to a vacuum environment. More details on the apparatus can be found in the reference.

A calcium standard liquid sample of 1 ppm concentration was used to create a calcium ion beam from the ICPMS. An argon and its compounds ions generated from an ICP was suppressed by injecting 0.8 ml/min of ammonia gas into a reaction cell. A trap parameter was $V_{rf} = 500$ V, $\Omega = 2\pi \times 4.1$ MHz. A vacuum degree of the trap chamber was 1×10^{-9} Torr. The 397 nm laser frequency was tuned on -160 MHz from the ^{40}Ca$^+$ resonance for Doppler cooling of loaded ions and the 866 nm laser frequency was tuned on its resonance. Injected ions from the ICPMS were loaded and cooled on these experimental conditions. After finishing the loading a typical cooling spectrum of ^{40}Ca$^+$ ions was obtained. During the 397 nm frequency scanning a laser induced fluorescence (LIF) gradually increased then a phase transition appeared at -125 MHz red detuned frequency from the resonance. When the 397 nm frequency was blue detuned from the resonance a sharp decrease of the LIF due to laser heating was appeared and finally a FWHM of the spectrum was measured as 27.9 MHz. From these results we can conclude the cooling lasers stabilized by the performance of the DFOC system was good enough to cool Ca$^+$ ions.

3.3 Laser frequency switch for even isotope ion cooling

As shown in Table 1, resonance frequency difference of each Ca$^+$ isotope ions is different up to around 8 GHz due to its isotope shift. Therefore, the 397 and 866 nm laser frequencies have to be simultaneously switched to the resonance frequencies of target ions for Doppler cooling. The system simultaneously controls the frequency of cooling and repumping lasers from one Ca$^+$ isotope resonance to the other one by referring the relative fringe position from the etalon. When the two isotope numbers were designated for the resonance frequency switching, the computer system calculates a difference of the isotope shift shown in Table 1 and then simultaneously switch the frequency of 397 and 866 nm laser. Negative feedback

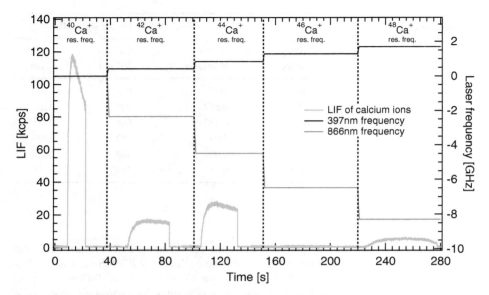

Fig. 4 The experimental result of a cooling laser frequency synchronization with the resonance of injected target ions from the ICPMS by the DFOC system. The 397 and 866 nm laser frequencies were switched from one even isotope to another one. The LIF from trapped and cooled ions was observed when the target ions from the ICPMS was reached at the trap segment

control technique [30] was adapted for all the slave lasers to increase a continuous scan range of our ECDL from 2 ∼ 3 GHz to 10 GHz.

A performance of the laser frequency switching function was evaluated by cooling stable even Ca^+ isotope ions in the ICPMS-ILECS. When the 397 and 866 nm laser frequencies were tuned on the resonance of a target isotope ion by the DFOC system then Ca^+ ion beam injected from the ICPMS was chosen in order of $^{40}Ca^+$, $^{42}Ca^+$, $^{44}Ca^+$, $^{46}Ca^+$ and $^{48}Ca^+$. The odd isotope $^{43}Ca^+$ was not considered here because its non-zero nuclear spin gives rise to the hyperfine structure and therefore its laser cooling scheme is different from that of even isotope ions. The experimental result was shown in Fig. 4. First, the 397 nm laser frequency was tuned on -170 MHz from the $^{40}Ca^+$ resonance and 866 nm frequency was tuned on the $^{40}Ca^+$ resonance to decrease a temperature of the loaded ions by Doppler cooling. 2.6×10^{-6} Torr of Helium buffer gas was injected in the trap chamber during the experiment to decelerate the ion beam from the ICPMS. When the $^{40}Ca^+$ ion beam was injected into the trap segment at 9 second, an LIF from the trapped and laser cooled ions was observed. Next the trapped $^{40}Ca^+$ ions were removed from the trap by making rf voltage to zero at 22 second. Then the frequencies of the 397 and 866 nm lasers were switched from the resonance of $^{40}Ca^+$ to that of $^{42}Ca^+$ at 38 second. When the $^{42}Ca^+$ ion beam was injected from the ICPMS, an LIF from trapped and cooled $^{42}Ca^+$ was observed at 53 second. Then the same process was repeated for $^{44}Ca^+$, $^{46}Ca^+$ and $^{48}Ca^+$. All the even stable Ca^+ isotope ions LIF was observed from the experiment and we can conclude that the DFOC system successfully switched the frequency of 397 and 866 nm lasers for Ca^+ isotopes ion cooling. The reason why the LIF of $^{46}Ca^+$ was very small is that the amount of $^{46}Ca^+$ ions which reached at the trap segment from the ICPMS was very small due to its low natural abundance of 0.004 %.

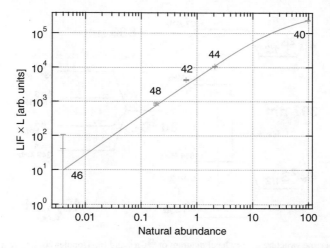

Fig. 5 A relationship between the natural abundance and a LIF $\times$ L value of even Ca$^+$ isotope ions which was trapped by a symmetrical control of the DFOC and the ICPMS

According to the ion loading model [23], the concentration C^χ of an isotope corresponds to a product of trapped isotope ions LIF and the ion loading rate L into the trap, where χ is a scaling factor. Therefore we have extracted the LIF maximum and loading rate information for each Ca$^+$ isotope from Fig. 4 to investigate the relationship between the trapped ions fluorescence and their natural abundance The Fig. 5 shows the calculated LIF $\times$ L and its linear fit. In fact the linear fitting worked well with $\chi = 0.88$ same as reference paper. The reason why $\chi < 1$ can be explained as follows. The amount of trapped ions were increased due to the high concentration of isotope, and therefore its ion cloud became large. With the large ion cloud, an effect of rf heating and micromotion became stronger and finally an ion temperature was increased. Consequently, an ion LIF was decreased with an isotope of which concentration was high. Moreover, a light collection efficiency of a detection system was also decreased due to the large ion cloud. And high error bar of ^{46}Ca$^+$ is because the 0.004 % of very rare natural abundance made the amount of trapped ions also very small and decreased the ion loading efficiency. From these results we can conclude lasers frequency controlled by the system were successfully synchronized to objective ions from the ICPMS Therefore Ca$^+$ isotopes ion loading and its selective cooling method was established. Furthermore, a correspondence between the LIF of the trapped isotope ions and its natural abundance was investigated.

3.4 Doppler cooling of odd isotope ion

The odd isotope ion cooling have to be considered differently from that of even isotopes due to its hyperfine structure. Calcium has a stable odd isotope ^{43}Ca natural abundance of which is 0.135 %. Therefore the ^{43}Ca$^+$ was chosen as a target ion here to evaluate the odd isotope ion cooling performance of the DFOC system. There are indeed several papers on ^{43}Ca$^+$ spectroscopy using ion trap such as a research using resonance ionization [31] or isotope concentrated environment [32]. Our laboratory reported the spectroscopy using sympathetic cooling [33] and selective heating and cooling method [34]. In this research we realized a direct ion loading environment which is different from the other researches by introducing

Springer

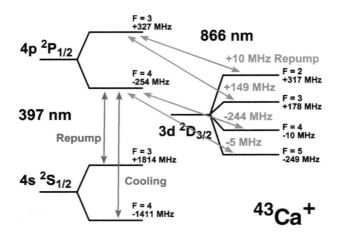

Fig. 6 The hyperfine structure energy level diagram of $^{43}Ca^+$ and the Doppler cooling scheme used in our experiment. Two 397 nm and four 866 nm lasers are required to close 4s $^2S_{1/2}$ ↔ 4p $^2P_{1/2}$ and 3d $^2D_{3/2}$ ↔ 4p $^2P_{1/2}$ transition. The beam which closing 4s $^2S_{1/2}$ (F=3) ↔ 4p $^2P_{1/2}$(F=4) transition was named as a repump 397 nm beam and cooling 397 nm beam for 4s $^2S_{1/2}$ (F=4) ↔ 4p $^2P_{1/2}$(F=4) transition

ICPMS. The hyperfine structure energy level diagram of $^{43}Ca^+$ and the Doppler cooling scheme we used are shown in Fig. 6. Although our DFOC system is available to control only two 397 nm and two 866 nm lasers, two 397 nm lasers are needed to close 4s $^2S_{1/2}$ ↔ 4p $^2P_{1/2}$ transition and four 866 nm lasers are needed to close 3d $^2D_{3/2}$ ↔ 4p $^2P_{1/2}$ transition as reported by the energy level diagram. However F:F'=5:4, 2:3 transition of 3d $^2D_{3/2}$ ↔ 4p $^2P_{1/2}$ can be closed with one 866 nm laser because the frequency difference between the two transition is sufficiently small. There is a report [35] about the 3d $^2D_{3/2}$ ↔ 4p $^2P_{1/2}$ transition of $^{43}Ca^+$ closing with only three 866 nm lasers. In the report a fundamental 866 nm laser was prepared to cover F:F'=3:3 and then created two modulated laser beams frequency of which are -150 MHz and -395 MHz from the fundamental laser beam to close F:F'=5:4, 2:3 and F:F'=4:4 respectively. Referring their 3d $^2D_{3/2}$ ↔ 4p $^2P_{1/2}$ repump system, we prepared the fundamental 866 nm laser beam covering F:F'=5:4, 2:3 and two modulated laser beams frequencies of which are +150 MHz for F:F'=3:3 and -245 MHz for F:F'=4:4 closing. With this 3d $^2D_{3/2}$ ↔ 4p $^2P_{1/2}$ hyperfine structure repump system a closed cooling cycle for $^{43}Ca^+$ with two 397 nm lasers and one 866 nm laser was able to be constructed and consequently all the frequencies of the lasers could be stabilized and controlled by the DFOC system. To evaluate an odd isotope ion cooling performance of the DFOC system, $^{43}Ca^+$ ions were trapped and cooled in the ICPMS-ILECS and its hyperfine splitting was also measured.

The frequencies of two 397 nm lasers were tuned on -200 MHz from each resonance and three 866 nm laser beams frequency was tuned on resonance for Doppler cooling of $^{43}Ca^+$ ions. All the five laser beams were adjusted to cross at the trap center Then $^{43}Ca^+$ ions which were created from a calcium standard liquid sample by the ICPMS were loaded in the trap and finally an LIF from the Doppler cooled $^{43}Ca^+$ ions was observed. To obtain a cooling spectrum of the trapped $^{43}Ca^+$, all the three 866 nm beams frequency were stabilized to repump 3d $^2D_{3/2}$ ↔ 4p $^2P_{1/2}$ transition and repump 397 nm laser frequency was also stabilized to close 4s $^2S_{1/2}$ (F=3) ↔ 4p $^2P_{1/2}$ (F'=4) transition (see Fig. 6). Then the frequency of cooling 397 nm laser was swept to observe a spectrum of 4s $^2S_{1/2}$ (F=4) ↔

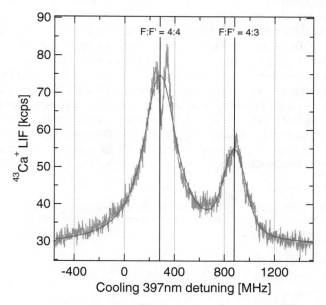

Fig. 7 4p $^2P_{1/2}$ level hyperfine splitting spectrum of ^{43}Ca$^+$. The splitting between F:F'=4:3 was measured as 597.2 MHz. The repump 397 nm laser has affected the amplitude difference of the two LIF peak and a dark resonance on the F=4 peak

4p $^2P_{1/2}$ transition. The 4p $^2P_{1/2}$ level cooling spectrum obtained from the trapped ^{43}Ca$^+$ ions is shown in Fig. 7. The x-axis frequency of the graph was calculated by converting the cooling 397 nm laser fringe movement using the etalon FSR and the LIF signal was fitted with a Voigt function. During the frequency sweeping two large peaks were observed at the F:F'=4:4 and F:F'=4:3 resonance. An amplitude of F:F'=4:4 peak was higher than the F:F'=4:3 peak in the graph We suppose the reason of LIF difference is the repump 397 nm laser. Ions in 4s $^2S_{1/2}$ (F=3) level were only excited to 4p $^2P_{1/2}$ (F'=4) level but not 4p $^2P_{1/2}$ (F'=3) level by the repump 397 nm laser Therefore the F'=4 level population of 4p $^2P_{1/2}$ was larger than that of F'=3 level during the experiment A strong dip in the F:F'=4:4 peak is a dark resonance resulted by the repump 397 nm laser.

The 4p $^2P_{1/2}$ level hyperfine splitting of ^{43}Ca$^+$ was measured as 597.2 MHz from the fitting and the reference value [36] is 581.7 MHz. The ion temperatures derived from the fitting to the Voigt function were 14 K for the for F:F'=4:4 transition and 7 K for the F:F'=4:3 transition, respectively. Two reasons can be think of the frequency difference between the measured hyperfine splitting and the literature value. First a Voigt function fitting was affected from the strong dark resonance dip in F:F'=4:4 transition therefore the peaks position would be deviated from a real position. Second a fluctuating LIF signal caused by the cooling 397 nm laser linewidth made the fitting error.

4 Conclusions

Laser frequencies were controlled for cooling stable calcium ions by the DFOC system using a computer and an etalon. The master laser was prepared by stabilizing lab-made 780 nm laser using Sub-Doppler DAVLL. The cavity length of the lab-made etalon was measured and stabilized to compensate its variation by environment condition change Four slave lasers for ion cooling were stabilized simultaneously. The ion cooling performance of the DFOC system was evaluated by observing the LIF of cooled $^{40}Ca^+$ ions. The frequency shift function was implemented to correspond the isotope shifts of target ions and all the stable even Ca^+ isotope ions were cooled and observed. Finally cooling laser system for the odd isotope ion with the hyperfine structure was constructed and controlled by the DFOC system. A cooling spectrum of $^{43}Ca^+$ ions was obtained and its hyperfine splitting was measured We are considering to introduce a FPGA (Field-Programmable Gate Array) control system and a high speed scanning etalon to narrow the linewidth of controlled lasers using its wide feedback bandwidth for better experiment precision as a next step of this research.

Acknowledgments The authors appreciate the technical assistance and valuable discussions with Dr. Miyabe in Japan Atomic Energy Agency And also thank him for generously providing the spectrum analyzer for our research.

References

1. Wieman, C.E., Hollberg, L.: Rev. Sci. Instrum. **62**, 1 (1991)
2. Galbács, G.: Appl. Spectrosc. Rev. **41**, 259 (2006)
3. Liger, V., Zybin, A., Kuritsyn, Y., Niemax, K.: Spectrochim. Acta, Part B **52**, 1125 (1997)
4. Dahmani, B., Hollberg, L., Drullinger, R.: Opt. Lett. **12**, 876 (1987)
5. Littman, M.G.: Opt. Lett. **3**, 138 (1978)
6. Ricci, L., Weidemüller, M., Esslinger, T., Hemmerich, A., Zimmermann, C., Vuletic, V., König, W., Hänsch, T.W.: Opt. Commun. **117**, 541 (1995)
7. Li, R., Jia, S., Loe-Mie, R., Bloch, D., Ducloy, M.: Laser Phys. **8**, 670 (1998)
8. Tiwari, V.B., Singh, S., Mishra, S.R., Rawat, H.S., Mehendale, S.C.: Opt. Commun. **263**, 249 (2006)
9. Corwin, K.L., Lu, Z.-T., Hand, C.F., Epstein, R.J., Wieman, C.E.: Appl. Opt. **37**, 3295 (1998)
10. Wasik, G., Gawlik, W., Zachorowski, J., Zawadzki, W.: Appl. Phys. B Lasers Opt. **75**, 613 (2002)
11. Helmcke, J., Lee, S.A., Hall, J.L.: Appl. Opt. **21**, 1686 (1982)
12. Drever, R.W.P., Hall, J.L., Kowalski, F.V., Hough, J., Ford, G.M., Munley, A.J., Ward, H.: Appl. Phys. B **31**, 97 (1983)
13. Zhu, M., Hall, J.L.: J. Opt. Soc. Am. B **10**, 802 (1993)
14. Marino, A.M., Jr, C.R.S.: Rev. Sci. Instrum. **79**, 013104 (2008)
15. Bushaw, B.A., Cannon, B.D., Gerke, G.K., Whitaker, T.J.: Opt. Lett. **11**, 422 (1986)
16. Zhao, W.Z., Simsarian, J.E., Orozco, L.A., Sprouse, G.D.: Rev. Sci. Instrum. **69**, 3737 (1998)
17. Lindsay, B.G., Smith, K.A., Dunning, F.B.: Rev. Sci. Instrum. **62**, 1656 (1991)
18. Riedle, E., Ashworth, S.H., Farrell, J.T., Nesbitt, D.J.: Rev. Sci. Instrum. **65**, 42 (1994)
19. Matsubara, K., Uetake, S., Ito, H., Li, Y., Hayasaka, K., Hosokawa, M.: Jpn. J. Appl. Phys. **44**, 229 (2005)
20. Seymour-Smith, N., Blythe, P., Keller, M., Lange, W.: Rev. Sci. Instrum. **81**, 075109 (2010)
21. Jaffe, S.M., Rochon, M., Yen, W.M.: Rev. Sci. Instrum. **64**, 2475 (1993)
22. Zhao, W.Z., Simsarian, J.E., Orozco, L.A., Sprouse, G.D.: Rev. Sci. Instrum. **69**, 3737 (1998)
23. Kitaoka, M., Jung, K., Yamamoto, Y., Yoshida, T., Hasegawa, S.: J. Anal. At. Spectrom **28**, 1648 (2013)
24. Miyabe, M., Oba, M., Kato, M., Wakaida, I., Watanabe, K.: J. Nucl. Sci. Technol. **43**, 305 (2006)
25. Marciante, M., Champenois, C., Calisti, A., Pedregosa-Gutierrez, J., Knoop, M.: Phys. Rev. A **82**, 033406 (2010)
26. Jin, J., Church, D.A.: Phys. Rev. A **49**, 3463 (1994)
27. Nörtershäuser, W., Blaum, K., Icker, K., Müller, P., Schmitt, A., Wendt, K., Wiche, B.: Eur. Phys. J. D Atom. Mol. Opt. Phys. **2**, 33 (1998)
28. Petelski, T., Fattori, M., Lamporesi, G., Stuhler, J., Tino, G.M.: Eur. Phys. J. D **22**, 5 (2003)
29. Masabumi, M., Masaaki, K., Masaki, O., Ikuo, W., Kazuo, W., Klaus, W.: Jpn. J. Appl. Phys. Part 1 **45**, 4120 (2006)
30. Nayuki, T., Fujii, T., Nemoto, K., Kozuma, M., Kourogi, M., Ohtsu, M.: Opt. Rev. **5**, 267 (1998)

31. Lucas, D.M., Ramos, A., Home, J.P., McDonnell, M.J., Nakayama, S., Stacey, J.-P., Webster, S.C., Stacey, D.N., Steane, A.M.: Phys. Rev. A **69**, 012711 (2004)
32. Kurth, F., Gudjons, T., Hilbert, B., Reisinger, T., Werth, G., Mårtensson-Pendrill, A.-M.: Z. Phys. D: At., Mol. Clusters **34**, 227 (1995)
33. Hashimoto, Y., Kitaoka, M., Yoshida, T., Hasegawa, S.: Appl. Phys. B Lasers Opt. **103**, 339 (2011)
34. Kitaoka, M., Hasegawa, S.: J. Phys. B Atomic Mol. Phys. **45**, 165008 (2012)
35. Benhelm, J., Kirchmair, G., Roos, C.F., Blatt, R.: Phys. Rev. A **77**, 062306 (2008)
36. Sahoo, B.K.: Rev, Phys. A **80**, 012515 (2009)

Hyperfine Interact (2015) 236:53–58
DOI 10.1007/s10751-015-1193-1

Magneto-optical trapping of radioactive atoms for test of the fundamental symmetries

Hirokazu Kawamura[1,2] · S. Ando[2] · T. Aoki[2] · H. Arikawa[2] · K. Harada[2] ·
T. Hayamizu[2] · T. Inoue[1,2] · T. Ishikawa[2] · M. Itoh[2] · K. Kato[2] · L. Köhler[2] ·
J. Mathis[2] · K. Sakamoto[2] · A. Uchiyama[2] · Y. Sakemi[2]

Published online: 22 May 2015
© Springer International Publishing Switzerland 2015

Abstract We are planning test experiments of fundamental symmetries based on the intrinsic properties of francium. It is expected that the laser cooling and trapping of francium will produce precision measurements. The pilot experiment using rubidium was performed with the goal of francium trapping. The ion beam generated with a francium ion source was investigated using a Wien filter. Each piece of equipment still must be studied in more detail, and the equipment should be upgraded in order to trap radioactive atoms.

Keywords Francium · Magneto-optical trap · Wien filter · Standard model

1 Introduction

Francium (Fr) is one of the newest elements and the heaviest element among alkali metals. Fr is a radioactive element including many isotopes with mass numbers 199 to 232; they are never stable isotopes. The group ISOLDE-CERN succeeded in finding and measuring the optical transition of online-produced Fr isotopes in the 1970s [1]. In 1996, a group at the State University of New York achieved the magneto-optical trapping of Fr [2]. It is believed that Fr is suited for use in searching for the electric dipole moment of the electron and for the nuclear anapole moment [3].

Proceedings of the 6th International Conference on Trapped Charged Particles and Fundamental Physics (TCP 2014), Takamatsu, Japan, 1–5 December 2014.

✉ Hirokazu Kawamura
  hirokazu.kawamura.c2@tohoku.ac.jp

1   Frontier Research Institute for Interdisciplinary Sciences, Tohoku University, Sendai, Miyagi
    980-8578, Japan

2   Cyclotron and Radioisotope Center, Tohoku University, Sendai, Miyagi 980-8578, Japan

The electric dipole moment (EDM) of the electron is observable, immediately indicating violation of the time-reversal symmetry. A search for the EDM entails a test of the standard model and a search for the new physics beyond the standard model, since the time-reversal symmetry is held within the framework of the standard model. Many groups have searched for the EDM using many methods in many systems. Measurement precision has continued to improve, though the reported values are upper limit. According to the Schiff theorem, the atomic EDM cannot be observed because the subatomic EDM is screened in neutral atoms [4]. In paramagnetic atoms, the Schiff screening is violated by the relativistic effect. The unpaired electron in the outermost shell experiences the nuclear electric field, and the EDM effect will appear with an enhancement. Since the enhancement effect increases with the nuclear charge, the electron EDM experiment was performed using heavy atoms with an unpaired electron, such as cesium [5] or thallium [6].

The parity non-conservation effect is well known in atomic systems. The main component of the atomic parity violation is independent of the nuclear spin. The nuclear-spin dependent effect, which is much smaller than the independent effect, predominantly originates from the nuclear anapole moment (AM). The AM originates from the weak interaction between the nucleons and increases with the number of nucleons. A simple atomic structure is preferable in order to accurately extract the AM effect from the atomic parity-violating effect. Therefore, experiments have used cesium [7, 8] and thallium [9, 10], which are relatively heavy and simple. In 1997, a cesium experiment reported that the AM effect was observed [7]. This measurement aligns well with the theory in terms of the nuclear-spin independent parity violation [11]. However, some measurements conflict with the results of the spin-dependent parity violation in this cesium experiment [3]. Further experiments are needed in order to explore this contradiction.

1.1 Precision measurements with francium

Atoms with a large nucleus and simple electronic structure are favorable for investigating the electron EDM and nuclear AM. Fr is one of the most suitable atoms meeting these conditions. However, statistical precision tends to be a problem due to the absence of a stable isotope. The application of laser cooling and trapping to Fr will overcome this problem. An electrostatic field must be applied to atoms in order to measure these moments. The period that the atoms interact with the external field is too short ($\sim$msec) in atomic beam experiments. If the atoms are trapped, the interaction time can be longer ($\sim$sec) and will cover the paucity of the objective atoms. On the basis of this motivation, precision measurement of the symmetry violation is planned using Fr. Some groups have already performed the laser trapping experiment of Fr at TRIUMF [12] and LNL [13]. At the Cyclotron and Radioisotope Center at Tohoku University, development experiments also are working toward Fr trapping [14].

We planned the magneto-optical trapping of Fr as follows: Fr is produced through the nuclear fusion reaction between an oxygen beam and gold target. The Fr produced inside the target is diffused and desorbed by heat because the gold is highly heated. Electrostatic fields extract the ionized Fr from the desorbed particles. The fusion reaction produces not only Fr but also radiation, such as neutron and γ rays, which would interfere with the precision measurement. Hence, the extracted ions are transported into the next room, shielded from the radiation. The transported ions stop at the surface of a yttrium target. The yttrium neutralizer is then heated and desorbs the Fr as neutral atoms. Finally, the desorbed neutral Fr atoms are captured in a vacuum by applying laser beams and quadrupole

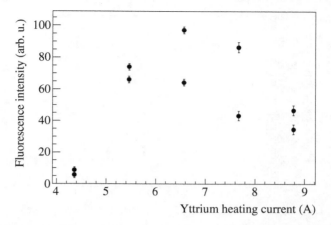

Fig. 1 Dependence on the temperature of yttrium neutralizer target for the fluorescence intensity. The *vertical axis* shows the intensity of the fluorescence emitted from the trapped atoms that corresponds to the number of trapped atoms. The *horizontal axis* shows the electric current heating the yttrium neutralizer that corresponds to the yttrium temperature. The current of 7 A would be roughly 700 °C

magnetic field. This proceeding reports the developmental status of each experimental component.

2 Experiment

2.1 Trapping experiment using rubidium

Rubidium (Rb), which is an alkali atom similar to Fr, is suitable for equipment development. We have already achieved the ion production, transportation, neutralization and magneto-optical trap of a stable Rb isotope. The trapping efficiency of the magneto-optical trap depends on how the neutralized atoms are emitted. The number of trapped atoms is investigated in terms of the dependence on the neutralizer conditions.

As the temperature of the neutralizer target gets higher, the atomic desorption is enhanced, but the trapping efficiency worsens because of a higher velocity. Therefore, there should be an optimum temperature to maximize the number of trapped atoms. Figure 1 shows the target temperature dependence of the trapped atoms. As expected, we obtained the result indicating the existence of the optimum temperature.

The trapping efficiency should depend on the energy of the incident ion beam. Particles become difficult to desorb as a higher energy beam implants deeper into the neutralizer. The beam energy dependence of the trapped atoms is shown in Fig. 2. The lower-beam energy leads to an increased number of trapped atoms. In fact, it has been found empirically that reducing the beam energy (i.e., the acceleration voltage) results in a decrease in the ion extraction efficiency and transport efficiency. The energy dependence of the ion beam current as well as the trapping efficiency must be considered in order to maximize the trapped atoms.

At optimum conditions, the number of trapped Rb atoms was roughly estimated to be 10^5 atoms. Since the number of Rb ions accumulated on the neutralizer was 10^9 ions/sec $\times$ 10 sec $= 10^{10}$ ions, the trapping efficiency would be approximately 10^{-5} as a rough order estimation.

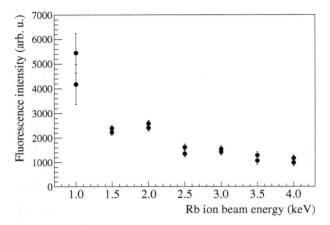

Fig. 2 The beam energy dependence of the number of trapped atoms. The fluorescence intensity, which corresponds to the number of atoms, is normalized by the ion beam current

2.2 Purification of the ion beam

When the Fr desorbs and ionizes at the surface of the gold target, it is expected that other particles become ion beams through the same process as Fr. The number of ^{210}Fr can be estimated by measuring the number of α particles using a solid-state detector. Our ^{210}Fr beam intensity is typically 10^5 pps, roughly corresponding to the current of 10^{-5} nA. On the other hand, the beam current is typically measured at 10 nA with a Faraday cup. This large current is barely sensitive to the primary oxygen beam. A large amount of background component induces frequent atomic collisions and subsequently disturbs the laser trapping. A pure ion beam is required for efficient trapping.

A Wien filter was included in the beam transport system to purify the ion beam. If all ions have the same energy, the components other than Fr can be separated using the Wien filter, which acts as a mass separator. The mass spectrum obtained using the filter is shown in Fig. 3. The mass-charge ratio is determined by the relationship between the applied field intensity and the mass. According to the preceding study [15], clear peaks would derive from alkali metals such as sodium, potassium and rubidium, which are impurities in the material. A broad peak was found to be around 200 of the mass/charge, and this peak was also observed without the primary beam. A possible explanation is that the gold target itself is ionized. In principle, heavier particles are difficult to separate by a Wien filter; therefore, the peaks in the spectrum get broader as the mass increases. If the field intensity reaches 200 mT, which is within the original design, it will lead to a good mass resolution to separate Fr from Au.

The background current for Fr beam drops from 10 nA to 0.1 nA in this experimental condition (Fig. 3). In other words, this filter can improve the purity by at least 100 times as long as the transport efficiency of Fr ions does not change, regardless of whether the Wien filter is used. Our transport system has an instability which could originate from a charging of insulating parts. Because of this instability, transport parameters require frequent optimization and are difficult to reproduce. In such a situation, the transport efficiency, which is defined as the ratio of the beam intensity at the first detector and at the last detector, was 15 % without the filter and 26 % with the filter. It is possible that transport efficiency could

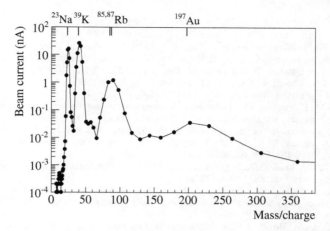

Fig. 3 Mass spectrum of the ion beam measured with the Wien filter. The magnetic field of the filter was fixed at approximately 40 mT. The electric voltage varied from 100 to 680 V across the 60 mm gap. The acceleration voltage of the beam was 3 kV. The measured temperature of the gold target was approximately 900 °C

be improved by a weak focusing effect of the Wien filter. Further investigation is necessary to discover how this purification leads to an increase in the number of trapped atoms.

3 Summary and plans

We are carrying out a project that traps radioactive francium atoms toward tests of fundamental symmetries through the measurement of the electric dipole moment and the anapole moment. Trapping properties have been studied earlier using stable rubidium atoms, and the ion beam was purified with a Wien filter. Other developments have also been considered, such as an upgrade of the Fr ion source and redesign of a trapping glass cell. In particular, the glass cell is strongly related to the trapping efficiency. Ideally, the distance from the neutralizer to the trapping area should be as short as possible, and the whole volume of the cell should be irradiated with laser beams, as the TRIUMF group has already shown. Our present cell, however, does not meet the ideal conditions at all because the required glassmaking is too difficult. The trapping efficiency will be improved when a highly-skilled glassmaker creates an ideal cell. The performance evaluation and upgrade of the experimental components must be executed in the facility to achieve the Fr trapping.

Acknowledgments　We gratefully acknowledge K. Hatanaka, A. Tamii, H.P. Yoshida and staffs of RCNP at Osaka University for their donation of the Wien filter.

Compliance with ethical standards

Funding　This work was supported by Grants-in-Aid for Scientific Research (KAKENHI), Grant Numbers 26220705 and 25610112, of the Japan Society for the Promotion of Science (JSPS). We thank JSPS Bilateral Joint Research Project between JSPS and INSA.

Conflict of interest　The authors declare that they have no conflicts of interest.

References

1. Liberman, S., Pinard, J., Duong, H.T., Juncar, P., Pillet, P., Vialle, J.-L., Jacquinot, P., Touchard, F., Büttgenbach, S., Thibault, C., de Saint-Simon, M., Klapisch, R., Pesnelle, A., Huber, G.: Laser optical spectroscopy on francium D_2 resonance line. Phys. Rev. A **22**, 2732–2737 (1980)
2. Simsarian, J.E., Ghosh, A., Gwinner, G., Orozco, L.A., Sprouse, G.D., Voytas, P.A.: Magneto-Optic Trapping of ^{210}Fr. Phys. Rev. Lett. **76**, 3522–3525 (1996)
3. Ginges, J.S.M., Flambaum, V.V.: Violations of fundamental symmetries in atoms and tests of unification theories of elementary particles. Phys. Rep. **397**, 63–154 (2004)
4. Schiff, L.I.: Measurability of nuclear electric dipole moments. Phys. Rev. **132**, 2194–2200 (1963)
5. Murthy, S.A., Krause, D. Jr.., Li, Z.L., Hunter, L.R.: New limits on the electron electric dipole moment from cesium. Phys. Rev. Lett. **63**, 965–968 (1989)
6. Regan, B.C., Commins, E.D., Schmidt, C.J., DeMille, D.: New limit on the electron electric dipole moment. Phys. Rev. Lett. **88**, 071805 (2002)
7. Wood, C.S., Bennett, S.C., Cho, D., Masterson, B.P., Roberts, J.L., Tanner, C.E., Wieman, C.E.: Measurement of parity nonconservation and an anapole moment in cesium. Science **275**, 1759–1763 (1997)
8. Guéna, J., Lintz, M., Bouchiat, M.A.: Measurement of the parity violating $6S$-$7S$ transition amplitude in cesium achieved within 2×10^{-13} atomic-unit accuracy by stimulated-emission detection. Phys. Rev. A **71**, 042108 (2005)
9. Edwards, N.H., Phipp, S.J., Baird, P.E.G., Nakayama, S.: Precise measurement of parity nonconserving optical rotation in atomic thallium. Phys. Rev. Lett. **74**, 2654–2657 (1995)
10. Vetter, P.A., Meekhof, D.M., Majumder, P.K., Lamoreaux, S.K., Fortson, E.N.: Precise test of electroweak theory from a new measurement of parity nonconservation in atomic thallium. Phys. Rev. Lett. **74**, 2658–2661 (1995)
11. Olive, K.A., et al.: (Particle Data Group), Review of particle physics. Chin. Phys. C **38**, 090001 (2014)
12. Tandecki, M., Zhang, J., Collister, R., Aubin, S., Behr, J.A., Gomez, E., Gwinner, G., Orozco, L.A., Pearson, M.R.: Comissioning of the francium trapping facility at TRIUMF. J. Instrum. **8**, P12006 (2013)
13. Mariotti, E., Khanbekyan, A., Marinelli, C., Marmugi, L., Moi, L., Corradi, L., Dainelli, A., Calabrese, R., Mazzoccca, G., Tomassetti, L.: Francium trapping at the INFN-LNL facility. Int. J. Mod. Phys. E **23**, 1430009 (2014)
14. Kawamura, H., Ando, S., Aoki, T., Arikawa, H., Ezure, S., Harada, K., Hayamizu, T., Inoue, T., Ishikawa, T., Itoh, M., Kato, K., Kato, T., Nataraj, H.S., Sato, T., Uchiyama, A., Aoki, T., Furukawa, T., Hatakeyama, A., Hatanaka, K., Imai, K., Murakami, T., Shimizu, Y., Wakasa, T., Yoshida, H.P., Sakemi, Y.: Search for a permanent EDM using laser cooled radioactive atom. EPJ Web Conf. **66**, 05009 (2014)
15. Sanguinetti, S.: PhD thesis, the Université Pierre et Marie Curie and the Università di Pisa, Atomic Parity Violation in Heavy Alkalis: Detection by Stimulated Emission for Cesium and Traps for Cold Francium, p. 126

Hyperfine Interact (2015) 236:59–64
DOI 10.1007/s10751-015-1203-3

Performance assessment of a new laser system for efficient spin exchange optical pumping in a spin maser measurement of ^{129}Xe EDM

C. Funayama[1] · T. Furukawa[2] · T. Sato[1] · Y. Ichikawa[1,3] · Y. Ohtomo[1] ·
Y. Sakamoto[1] · S. Kojima[1] · T. Suzuki[1] · C. Hirao[1] · M. Chikamori[1] ·
E. Hikota[1] · M. Tsuchiya[1] · A. Yoshimi[4] · C. P. Bidinosti[5] · T. Ino[6] · H. Ueno[3] ·
Y. Matsuo[7] · T. Fukuyama[8] · K. Asahi[1]

Published online: 26 October 2015
© Springer International Publishing Switzerland 2015

Abstract We demonstrate spin-exchange optical pumping of ^{129}Xe atoms with our newly made laser system. The new laser system was prepared to provide higher laser power required for the stable operation of spin maser oscillations in the ^{129}Xe EDM experiment. We studied the optimum cell temperature and pumping laser power to improve the degree of ^{129}Xe spin polarization. The best performance was achieved at the cell temperature of 100 °C with the presently available laser power of 1 W. The results show that a more intense

Proceedings of the 6th International Conference on Trapped Charged Particles and Fundamental Physics (TCP 2014), Takamatsu, Japan, 1–5 December 2014

✉ C. Funayama
funayama@yap.nucl.ap.titech.ac.jp

1 Department of Physics, Tokyo Institute of Technology, 2-12-1 Ookayama, Meguro, Tokyo 152-8551, Japan

2 Department of Physics, Tokyo Metropolitan University, 1-1 Minami-Osawa, Hachioji, Tokyo 192-0397, Japan

3 RIKEN Nishina Center, RIKEN, 2-1 Hirosawa, Wako, Saitama 351-0198, Japan

4 Research Core for Extreme Quantum World, Okayama University, 3-1-1 Tsushimanaka, Kita, Okayama 700-8530, Japan

5 Department of Physics, University of Winnipeg, 515 Portage Avenue, Winnipeg, Manitoba, Canada

6 High Energy Accelerator Research Organization (KEK), Institute of Material Structure Science, 1-1 Oho, Tsukuba, Ibaraki 305-0801, Japan

7 Department of Advanced Sciences, Hosei University, 3-7-2 Kajino-cho, Koganei, Tokyo 184-8584, Japan

8 Research Center for Nuclear Physics (RCNP), Osaka University, 10-1 Mihogaoka, Ibaraki, Osaka 567-0047, Japan

laser is required for further improvement of the spin polarization at higher cell temperatures in our experiment.

Keywords Spin-exchange optical pumping · Spin polarization in diamagnetic atoms

1 Introduction

A permanent electric dipole moment (EDM) violates time-reversal invariance, and hence serves as a key observable to test theories beyond the Standard Model. We aim to search for an EDM in diamagnetic ^{129}Xe atom beyond the present upper limit of 4.1×10^{-27} $e$cm [1], at the order of 10^{-28} $e$cm, using an active feedback spin maser technique that sustains the nuclear spin precession semi-permanently [2, 3]. In the maser experiment, the degree of atomic spin polarization is one of the important parameters because the polarization directly influences the stability of the maser operation and then the achieved precision in our measurement. The spin polarization of ^{129}Xe is produced by spin-exchange optical pumping (SEOP) using spin polarized Rb atoms [4]. Thus, both the degree of spin polarization in Rb atoms and the Rb-^{129}Xe spin-exchange rate are important for the efficiency of SEOP.

Recently we have installed a ^{3}He co-magnetometer to remove the systematic errors of the observed maser frequency caused by drift of magnetic field. An atomic EDM in ^{3}He is expected to be negligibly small because of the small atomic number. The maser oscillation of ^{3}He, however, turned out to be much more unstable compared to that of ^{129}Xe. We consider that the instability of ^{3}He maser oscillation originates from insufficient polarization of ^{3}He due to an inefficient spin exchange between Rb and ^{3}He, which is typically three orders of magnitude smaller than that between Rb and ^{129}Xe [5, 6]. An effective way to overcome the small spin exchange rates would be to increase the number of Rb atoms by increasing the cell temperature. However, the hotter the cell temperature becomes, the higher the laser power for the optical pumping must be. We therefore prepared a new self-made intense laser system.

2 New laser system

Our new laser system (TA-ECLD) includes a self-made, Littrow type external cavity laser diode (ECLD) as a seed laser and an intense tapered amplifier (TA). ECLD could have narrow line width and good frequency stability as compared to other lasers. In the ECLD we employ a laser diode (LD) with anti-reflection coating (Toptica Photonics, LD-0790-0120-AR-1, power: up to 120 mW). A gold-coating replicated holographic grating (Optometrics, 3-4182) is used to selectively reflect a small range of the LD's emission spectrum back into the LD to narrow the laser linewidth. The free-running wavelength is set at D1 resonance line of Rb atoms in the cell we used (794.980 ± 0.001 nm) by adjusting the grating angle. A GaAs based TA (Eagleyard Photonics, EYP-TPA-0795-02000-4006-CMT04-0000) is used for the amplification of the seed laser light from the ECLD. The ECLD and the TA are coupled as shown in Fig. 1. The seed light power of 40 mW was amplified to 2 W which is almost the diffraction limited power of TA. However the intensity of laser light obtained in the present system was up to 1 W at the SEOP cell. A main cause of the decrease of laser intensity is considered to be the reflectance and transmittance in the isolator and related optical elements.

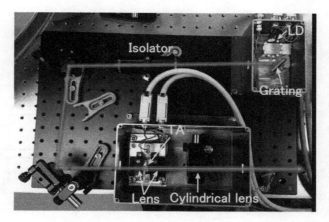

Fig. 1 Photograph of the fabricated TA-ECLD system. The grating is used to selectively reflect a small range of the LD's emission spectrum back into the LD to narrow the laser linewidth. The output of ECLD was amplified with TA. We used two lenses to collimate the output beam from TA, one was an aspheric lens (Thorlabs, C390TME-B) and the other a cylindrical lens (Thorlabs, LJ1695RM-B, focal length 50 mm), because the beam divergences for the horizontal and vertical direction were different. 2 W was obtained just below the cylindrical lens. Two isolators (Thorlabs, IOT-5-780-VLP, isolation: 55 dB) installed below the ECLD and the TA, respectively, were used to protect the laser diode from unwanted back reflections

3 Polarization measurement

The measurement was carried out for a spherical cell made of GE180 glass with a diameter of 20 mm. It contained approximately 1 Torr of ^{129}Xe, 425 Torr of ^{3}He and 100 Torr of N_2. The cell temperature was controlled with hot air from a heater. The degree of spin polarization was derived from the Adiabatic Fast Passage Nuclear Magnetic Resonance (AFP-NMR) signal (see Fig. 2). The nuclear spins of ^{129}Xe was placed under a static magnetic field produced with a Helmholtz coil, and were polarized by SEOP. The ^{129}Xe spin magnetization was flipped by the AFP-NMR with a RF field (frequency: 34.6 kHz) generated by a pair of small coils and the static magnetic field sweeping across the resonance at 29 G. Thus, the electromotive induction by the precession of nuclear magnetization was detected with another pair of coils (pick-up coils). The amplitude of detected signal was proportional to the magnetization. The picked-up signal was amplified with a lock-in amplifier (Stanford Research Systems, SR830).

The degree of spin polarization in ^{129}Xe, P_{Xe}, was deduced from the comparison of the observed NMR signals from ^{129}Xe and from protons in water whose polarization, P_p, was known from the Boltzmann distribution. Thus P_{Xe} is described as

$$P_{Xe} = \frac{g_p}{g_{Xe}} \times \frac{N_p}{N_{Xe}} \times \frac{V_{Xe}^{max}}{V_p^{max}} \times P_p \qquad (1)$$

where g_p and the g_{Xe} are the nuclear g factors for proton and ^{129}Xe, N_p and N_{Xe} are their number densities, and V_p^{max} and V_{Xe}^{max} are the measured amplitudes of their signals. In this study, we measured the polarizations of ^{129}Xe achieved at four cell temperatures of 80, 90, 100 and 110 °C and four laser powers of 0.2 W, 0.5 W, 0.7 W and 1.0 W. Note that the ^{129}Xe polarizations discussed in this paper are expressed relative to the polarization $P_{Xe}^{(0)}$

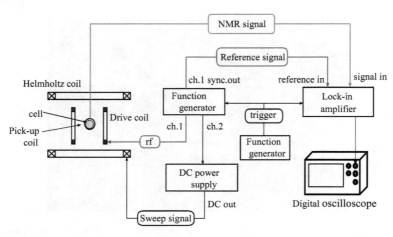

Fig. 2 Circuit diagram of the AFP-NMR setup [7]. A static magnetic field swept across the magnetic resonance at $B_0 = 29$ G with a RF frequency of 34.6 kHz. An induction signal in AFP-NMR was detected with the pick-up coils. The signal was amplified by a lock-in amplifier, recorded by a digital oscilloscope, and finally stored in a computer

that was obtained at 100 °C and 1 W, because the ^{129}Xe partial pressure in the cell had large uncertainty due to uncertain values of pressure determined during the cell preparation.

4 Result

Figures 3 and 4 show the ^{129}Xe relative spin polarization as a function of the laser power, and the cell temperature, respectively. Assuming the ^{129}Xe partial pressure to be 1 Torr, the polarization at 1 W and 100 °C was determined to be approximately 50 %. In the steady state, the ^{129}Xe spin polarization is described as

$$P_{Xe} = P_{Rb} \frac{\gamma_{se}}{\gamma_{se} + \Gamma_{sd}^{Xe}}, \tag{2}$$

where P_{Rb} is the Rb spin polarization, γ_{se} is the spin exchange rate between Rb and ^{129}Xe, Γ_{sd}^{Xe} is the spin relaxation rate of ^{129}Xe. The Rb spin polarization is represented as

$$P_{Rb} = \frac{\gamma_{opt}}{\gamma_{opt} + \Gamma_{sd}^{Rb}}, \tag{3}$$

where γ_{opt} is the optical pumping rate, which depends on the laser power, Γ_{sd}^{Rb} is the spin relaxation rate of Rb. The functional form used in the fitting analyses in Fig. 3 is $y = \frac{aX}{(X+b)}$ as suggested by (2) and (3). The spin polarization of ^{129}Xe is found to increase with the pumping laser power. The cell temperature of 100 °C is found to be the best temperature for ^{129}Xe within a range of laser power up to 1 W. The ^{129}Xe polarization at 110 °C does not seem to be saturated even at 1 W, indicating that the power was still not sufficient for such a dense Rb vapor (number density: 1.1×10^{13} cm^{-3}) [8]. The ^{129}Xe polarization attained for individual temperatures, which was deduced from the fitting, increases as the temperature becomes higher, because γ_{se} bears a proportionate relationship to Rb number density. However the ^{129}Xe polarization is lower at 110 °C than

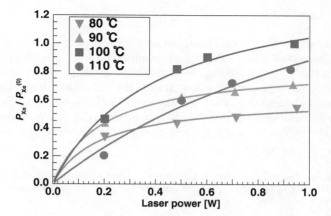

Fig. 3 Laser power dependence of the ^{129}Xe relative polarization for four different cell temperatures. The obtained polarizations were normalized with the polarization $P_{\mathrm{Xe}}^{(0)}$ at 1W and 100 °C. The data were fitted with a function $P_{\mathrm{Xe}}/P_{\mathrm{Xe}}^{(0)} = \frac{aX}{(X+b)}$, where a and b are constants to be determined through the fitting. Within a range of laser power up to 1 W, the polarization of ^{129}Xe does not saturate as the temperature becomes higher, suggesting that the Rb polarization deteriorates with the temperature

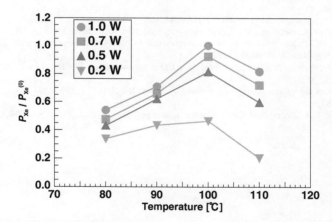

Fig. 4 Cell temperature dependence of the ^{129}Xe relative polarization for four different laser powers. Polarization at each temperature becomes higher as the laser power is increased. With the laser power fixed, the polarization first increases with the temperature, but starts to decrease above 100 °C, presumably because of the deterioration of Rb polarization, since the Rb vapor pressure rises up exponentially and, therefore, the optical pumping rate decreases due to excessive absorption of the laser light by Rb atoms and the spin relaxation rate of Rb increases

at 100 °C, since the degree of reduction in Rb spin polarization increases with increasing temperature. The cause is considered that Γ_{Rb} increases and γ_{opt} decreases due to enhanced laser light absorption by Rb atoms with elevated temperatures [9]. Based on these results, we expect that the spin polarization of Rb should be increased by increasing laser power.

 Springer

5 Summery and future

The TA-ECLD system was prepared and installed in the present work. Within a range of presently available laser powers, the highest polarization of ^{129}Xe is obtained with a cell temperature of 100 °C. In the next step, we plan to improve the laser power by incorporating an additional TA. The measurement will be made of the ^{3}He polarization at higher temperatures, the results of which should allow more quantitative discussions. The new laser system enables us to optimize the condition for maser oscillation, and thus to search for the best condition for the ^{129}Xe/^{3}He dual spin maser operation, in particular with maximized ^{3}He polarization.

Acknowledgments This work was partly supported by the JSPS KAKENHI (No.21104004, No.21244029 and No.26247036). One of the authors (T. Sato) would like to thank the JSPS Research Fellowships for Young Scientists for the support.

References

1. Rosenberry, M.A. et al.: Atomic electric dipole moment measurement using spin exchange pumped masers of ^{129}Xe and ^{3}He. Phys. Rev. Lett. **86**, 22 (2001)
2. Yoshimi, A. et al.: Nuclear spin maser with an artificial feedback mechanism. Phys. Lett. A **304**, 13 (2002)
3. Yoshimi, A. et al.: Low-frequency ^{129}Xe nuclear spin oscillator with optical spin detection. Phys. Lett. A **376**, 1924 (2012)
4. Happer, W.: Optical pumping. Rev. Mod. Phys. **44**, 169 (1972)
5. Coulter, K.P. et al.: Neutron polarization with polarized 3He. Nucl. Instr. Meth. A **270**, 90 (1988)
6. Cates, G.D. et al.: Rb-^{129}Xe spin-exchange rates due to binary and three-body collisions at high Xe pressures. Phys. Rev. A **45**, 4631 (1992)
7. Yoshimi, A.: Study of nuclear spin maser for EDM measurement. Ph. D. Thesis, Tokyo Institute of Technology (2000)
8. Alcock, C.B. et al.: Vapour pressure equations for the metallic elements: 298–2500 K. Can. Metall. Quart. **23**, 309 (1984)
9. Walker, T.G.: Fundamentals of spin-exchange optical pumping. J. Phys. Conf. Ser. **294**, 012001 (2001)

Hyperfine Interact (2015) 236:65–71
DOI 10.1007/s10751-015-1173-5

Extreme-field physics in Penning traps

The ARTEMIS and HILITE experiments

M. Vogel[1] · G. Birkl[2] · M. S. Ebrahimi[1] ·
D. von Lindenfels[1,3] · A. Martin[2] · G. G. Paulus[4] ·
W. Quint[1,3] · S. Ringleb[4] · Th. Stöhlker[1,4,5] · M. Wiesel[1,3]

Published online: 8 April 2015
© Springer International Publishing Switzerland 2015

Abstract We present two Penning trap experiments concerned with different aspects of the physics of extreme electromagnetic fields, the ARTEMIS experiment designed for bound-electron magnetic moment measurements in the presence of the extremely strong fields close to the nucleus of highly charged ions, and the HILITE experiment, in which well-defined ion targets are to be subjected to high-intensity laser fields.

Keywords Penning traps · Highly charged ions · Extreme fields

1 Introduction

The presence of extremely strong electromagnetic fields has a wide range of effects and may, amongst others, be studied in two interesting yet distinct regimes: on a microscopic scale there are extreme electric and magnetic fields in the vicinity of an atomic nucleus, which significantly alter the properties of bound electrons. As field strengths reach close to the Schwinger limit (of the order of 10^{16} V/cm, above which field production of real

Proceedings of the 6th International Conference on Trapped Charged Particles and Fundamental Physics (TCP 2014), Takamatsu, Japan, 1-5 December 2014

✉ M. Vogel
m.vogel@gsi.de

[1] GSI Helmholtzzentrum für Schwerionenforschung, 64291 Darmstadt, Germany

[2] Institut für Angewandte Physik, TU Darmstadt, 64289 Darmstadt, Germany

[3] Ruprecht Karls-Universität Heidelberg, Heidelberg, Germany

[4] IOQ, Universität Jena, 07743 Jena, Germany

[5] Helmholtz-Institut Jena, 07743 Jena, Germany

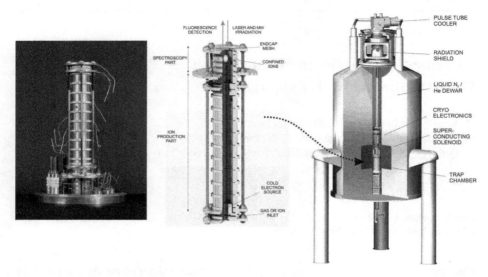

Fig. 1 Image of the ARTEMIS Penning trap (*left*), schematic of same trap (*middle*), and schematic of the complete setup with the trap located in the centre of the superconducting magnet (*right*)

electron-positron pairs would become possible), contributions from quantum electrodynamics (QED) play an important role to electronic structure, state lifetimes, and magnetic moments, and corresponding calculations can be tested with high accuracies [1]. In turn, this allows access to fundamental constants and symmetries. The ARTEMIS experiment is designed for precision measurements of bound electron magnetic moments in confined and cooled highly charged ions, and aims at measurements on the part per billion (ppb) level of accuracy for electronic *g*-factors and on the part per million (ppm) level of accuracy for nuclear magnetic moments. It is further designed for dedicated measurements of higher-order Zeeman effects. It is located at the HITRAP facility [2] at GSI, Germany, for access to low-energy highly charged ions. We present the concept, status, and first results.

On a macroscopic scale, extreme fields are present when atoms and ions are subjected to highly intense laser light. The electromagnetic fields in and close to a laser focus produce strongly non-linear optical effects such as multi-photon ionization to high charge states. The HILITE experiment hence features a Penning trap for the preparation and positioning of well-defined ion targets, as well as for non-destructive detection and confinement of reaction products in studies with various high-intensity and / or high-energy lasers.

2 Double-resonance spectroscopy at ARTEMIS

Precise measurements of fine structure and hyperfine structure transitions in highly charged ions allow sensitive tests of corresponding calculations in the framework of quantum electrodynamics of bound states [3]. Precisely measurable quantities comprise magnetic dipole (M1) transition energies and to some extent also lifetimes in the optical and in the microwave domain [4].

The obtainable spectroscopic resolution depends crucially on effects of line shift and broadening, prominently on first-order Doppler effects, which need to be minimized by

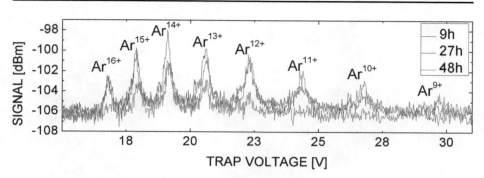

Fig. 2 Spectrum of in-trap created and confined argon ion charge states as a function of time upon creation

phase-space cooling of the ions' motions. To this end, ARTEMIS features techniques for extended ion storage and cooling prior to spectroscopic measurements.

The ARTEMIS experiment applies a laser-microwave double-resonance spectroscopy scheme which allows to precisely measure the Zeeman substructure of the fine or hyperfine structure of the ion under consideration [5]. From this, the magnetic moments (g-factors) of bound electrons can be determined for ions with non-zero nuclear spin, with precisions on the ppb scale, and in a somewhat complimentary approach to the Stern-Gerlach type measurements which have been successfully performed with various hydrogen-like ions [6–9]. At the same time, this information yields the nuclear magnetic moments with precisions on the ppm scale. A nice feature when applied to few-electron systems is the absence of diamagnetic shielding of the nucleus by outer electrons, hence these measurements enable benchmarks of shielding models. The principle of the laser-microwave double-resonance technique is to use fluorescence light from a closed optical transition as a probe for the microwave excitation between corresponding Zeeman sublevels. Different level schemes allow different preparation and measurement procedures, as has been discussed in detail in [5]. The envisaged experimental resolution allows to measure also quadratic and cubic contributions to the Zeeman effect, as has been detailed out in [10], and which allow the first laboratory access to individual higher-order contributions on the magnetic sector. The Penning trap in use for this kind of spectroscopy is a dedicated development to the end of maximizing the optical fluorescence yield, a so-called 'half-open' Penning trap [11]. It also features an ion creation part in full similarity to a cryogenic mini-EBIS [12], in which test ions such as Ar^{13+} are produced. This part of the trap is designed for dynamic capture of ions from an external source like an EBIS or from the HITRAP facility [2] via a low-energy beamline [14, 15]. We are currently commissioning the system with internally produced ions, see the charge spectrum in Fig. 2. It shows first measurements of the axial detection signal of the same ion cloud at three different storage times after ion creation. In the present spectrum, the signal is picked up by a resonant circuit at a frequency of $\omega = 2\pi \cdot 635$ kHz while the trap voltage U is ramped from 32 V to 16 V with a scanning speed of $dU/(Udt) \approx 10^{-2}$/s. From the observed space charge shift of the ion signal, we estimate an ion number density of about 10^6/cm^3 which is roughly one third of the expected electric space charge limit for that trap. Since three spectra of the same ion cloud are taken at different times, one can estimate the residual gas pressure. Assuming a cross section for electron capture from residual gas of $3.25 \cdot 10^{-15}$cm^2 for ions like Ar^{13+} [13], the observed charge state lifetime of roughly 20 hours results in a value for the residual gas pressure of

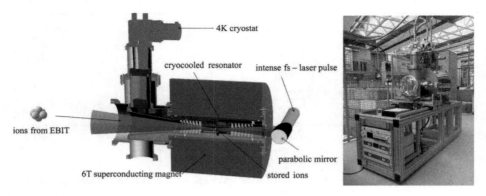

Fig. 3 Schematic of the HILITE experiment (*left*) and image of the actual setup (*right*)

about 10^{-13} hPa, which will allow storage of the highest charge states from the HITRAP facility for several hours, sufficient for the envisaged measurements.

3 High-intensity-laser reaction studies with HILITE

An important effect of high electromagnetic field strengths in atomic physics is non-linear ionization. Laser ionization is a widely investigated topic and there are several experiments and theories concerned with the high-energy [16–18] and high-intensity [19–21] photo-ionization regimes. For the sake of a clean reaction environment, it is desirable to work with ion targets prepared in a well-defined state concerning the charge distribution, position, shape, species and spatial density. To this end, we have conceived and built a dedicated Penning trap setup for ion-target preparation and non-destructive detection. The setup is designed to be compact (see Fig. 3), such that it may be moved readily to laser facilities such as FLASH [22], PHELIX [23], JETI 200 and POLARIS [24] or others, which cover a broad range of possible ionization parameters.

3.1 Experimental setup

The HILITE Penning trap is located inside the bore of a horizontal superconducting magnet with a maximum magnetic field of 6 T. The trap is a mechanically compensated open-endcap Penning trap [25, 26] which consists of an eight-fold segmented ring electrode and a pair of endcap electrodes with a separation of $2z_0 = 17.4$ mm. For dynamic capture of ions, additional electrodes are mounted on either end of the trap, which have a conical opening to accept laser beams up to an aperture of f/5. The inner trap diameter is 20.0 mm and hence the experiment is transparent for non-focused (laser) beams up to that diameter. The main concern of the experiment is to be able to prepare ion targets for laser irradiation and non-destructively detect the reaction products. To that end, a number of trap-specific techniques are combined which shall briefly be described.

3.2 Ion selection, cooling and positioning

Typically, ion ensembles which are produced inside a trap or captured from external sources may contain different atomic or molecular constituents in different charge states. Since each ion species of a certain charge-to-mass ratio q/m has a specific axial oscillation frequency in the trap, it is possible to excite any species selectively. The so-called SWIFT-technique (Stored Waveform Inverse Fourier Transform) [27] allows to excite any combination of q/m-regions resonantly and simultaneously. If the excitation amplitude is sufficiently large, the unwanted ions are resonantly ejected from the trap, leaving a cloud which consists of desired ion species only. The phase space of this remaining ion cloud can be cooled by resistive or sympathetic cooling, as detailed out in [28]. Axial positioning of the ion cloud as a whole is possible by use of a trap voltage asymmetry, i.e. an effective non-zero voltage across the endcaps. In this way, the centre of the ion oscillation is shifted with respect to the geometrical trap centre, and can be positioned with a resolution on the micrometer scale with respect to the laser focus. Details about the possibilities and requirements have been given in [29].

3.3 Ion target density and shape

To the end of defining the spatial density of the ion cloud, we employ the so-called 'rotating wall technique' [30–32]. In combination with a choice of the axial trapping potential, it can further be used to define the aspect ratio of the ion cloud. The ion cloud, always being an ellipsoid of rotation under these circumstances, may be deformed continuously from an oblate form (flat disc perpendicular to the central trap axis), to a spheroid and further to a prolate form (cigar shape along the central trap axis). To this end, a rotating dipole field is created by phase-shifted sinusoidal rf-signals applied to opposing segments of the ring electrodes. It produces a torque on the ion cloud as a whole and forces the global rotation frequency of the ion cloud to the revolution frequency ω_r of the rf-drive. This frequency uniquely determines the density n of the ion cloud, which has its maximum n_{max} at ω_r equal to half the free cyclotron frequency of the ions. The density n_{max} is determined by the values of the confining field strengths, and fundamentally limited to

$$n_{max} = \frac{\epsilon_0 \cdot B^2}{2m},$$ (1)

the so-called 'Brillouin-limit' [33, 34]. The maximum ion number density for Xe^+ ($m = 132 \, \text{u}$) at a magnetic field strength of $B = 6 \, \text{T}$ is about $7 \times 10^5 \, \text{mm}^{-3}$.

3.4 Ion detection and ion counting

For spectrometry of the confined ions species we use the FT-ICR (Fourier Transform Ion Cyclotron Resonance) technique. It relies on a pick-up of image currents created by the ion motion, analyzed by Fourier transformation to yield information on the charge-to-mass spectrum of the trap content and the relative ion numbers [35]. As such, it is a non-destructive detection method which keeps the analysed ions confined in the trap. We detect ions by broadband analysis of the induced currents inside ring electrodes as well as inside endcap electrodes for the radial and the axial ion oscillations, respectively. The currents are pre-amplified by a low-noise cryogenic amplifier which works at 4.2 K. Additionally, we employ three different resonators directly attached to the trap electrodes to achieve a higher signal for certain frequencies. The first is a helical resonator, which amplifies multiples of a fundamental frequency, so that all charge states of one species can be measured simultaneously with similar and high sensitivity. The second and third resonators are RLC-circuits, which are optimized for axial frequencies. The resonant frequencies are 229 kHz and 702 kHz, which are chosen to cover the detection of all charge states of all ions up to xenon within the range of possible trap voltages. Additionally, we intend to employ low-noise high-sensitivity charge amplifiers on electrodes at either side of the trap for charge counting of (educt) ions entering and (product) ions leaving the trap. These are to be gauged by destructive ion counters.

3.5 Laser ionization

In general, there are two regimes for the ionization with high-intensity photon fields, multiphoton ionization (MPI) and field ionization (FI), which are distinguished by their Keldysh parameter [36]. For high-power lasers with visible or near-infrared (NIR) radiation, field ionization dominates, and ionization probabilities can be calculated from experimental parameters [38]. One envisaged experiment with the HILITE setup uses a laser beam focused by an off-axis parabolic mirror with a silver coating through a laser window into the trap vacuum. We have chosen a fused silica window due to its high purity and high damage threshold. For lasers up to a peak intensity of 0.1 TW at an e^{-2}-beam diameter of 35 mm, non-linearities of fused silica can be neglected. The focus parameters are optimized for a maximum yield of high charge states. Typically, we can store up to 10^5 ions in an ion cloud with an axial extension of about 500 μm and a diameter of about 125 μm. Following [37], we have calculated the beam shape as well as the intensity distribution inside the focal volume assuming a laser beam quality factor of $M^2 = 1.6$ (spatial profile close to Gaussian) and an f-number of f/16. Using this spatial laser intensity distribution, for a pulse energy of 10 mJ at a pulse duration of 40 fs, we expect about 4000 stored Ar^+ ions to become ionized up to a Ar^{8+} [38]. For low charge states, all particles along the laser axis can be ionized easily and the number of ionized particles is limited by the geometries of ion cloud and laser beam. Product ions remain stored in the trap for further studies or further ionization. Currently, HILITE is being prepared for initial tests to be performed with ions from an external electron beam ion source (EBIS) and an offline test laser.

4 Summary

The ARTEMIS Penning trap experiment located at the HITRAP facility at GSI, Germany, is currently being commissioned with in-trap produced test ions like Ar^{13+}. In a next step, laser-microwave double-resonance experiments within the fine structure and its Zeeman sublevels of this ion will be performed, to the end of measuring the magnetic moment of the electron bound in a boron-like ion. At a later stage, similar measurements with ions of higher charge states up to U^{91+} from the HITRAP facility are foreseen. The HILITE experiment, currently under construction at the same site, will be tested with offline ions from the HITRAP low-energy beamline and an offline laser before being operated at high-intensity laser facilities.

Conflict of interests The authors declare that they have no conflict of interest.

References

1. Beier, T.: Phys. Rep. **339**, 79 (2000)
2. Kluge, H.-J., et al.: Adv. Quantum Chem. **53**, 83 (2007)
3. Vogel, M., Quint, W.: Ann. Phys. **525**, 505 (2013)

4. Vogel, M., Quint, W.: Phys. Rep. **490**, 1 (2010)
5. Quint, W., Moskovkin, D.L., Shabaev, V.M., Vogel, M.: Phys. Rev. A **78**, 032517 (2008)
6. Häffner, H., et al.: Phys. Rev. Lett. **85**, 5308 (2000)
7. Verdú, J., et al.: Phys. Rev. Lett. **92**, 093002 (2004)
8. Sturm, S., et al.: Phys. Rev. Lett. **107**, 023002 (2011)
9. Wagner, A., et al.: Phys. Rev. Lett. **110**, 033003 (2013)
10. von Lindenfels D., et al.: Phys. Rev. A **87**, 023412 (2013)
11. von Lindenfels, D., et al.: Hyp. Int. **227**, 197 (2014)
12. Alonso J., et al.: Rev. Sci. Instr. **77**, 03A901 (2006)
13. Mann, R.: Z. Phys. D **3**, 85 (1986)
14. von Lindenfels, D., et al.: Can. J. Phys. **89**, 79 (2011)
15. Andelkovic, Z., et al.: submitted to Nucl. Inst. Meth A (2015)
16. Richter, M., et al.: Phys. Rev. Lett. **102**, 163002 (2009)
17. Feldhaus, J., et al.: J. Phys. B **46**, 164002 (2013)
18. Guichard, R., et al.: J. Phys. B **46**, 164025 (2013)
19. Augst, S., et al.: Phys. Rev. Lett. **63**(20), 2212 (1989)
20. Becker, W., Liu, X., Ho, P., Eberly, J.H.: Rev. Mod. Phys. **84**, 1011 (2012)
21. Palaniyappan, S., et al.: J. Phys. B **39**, S357 (2006)
22. Ackermann, W., et al.: Nat. Photonics **1**, 336 (2007)
23. Bagnoud, V., et al.: Appl. Phys. B **100**, 137 (2010)
24. Kessler, A., et al.: Opt. Lett. **39**, 1333 (2014)
25. Gabrielse, G., Mackintosh, F.: Int. J. Mass Spectrom. Ion Proc. **57**, 1 (1984)
26. Gabrielse, G., Haarsma, L., Rolston, S.L.: Int. J. Mass Spectr. Ion Proc. **88**, 319 (1989)
27. Guan, S., Marshall, A.G.: Anal. Chem. **65**, 1288 (1993)
28. Vogel, M., et al.: Phys. Rev. A **90**, 043412 (2014)
29. Vogel, M., Quint, W., Paulus, G., Stöhlker, T.: Nucl. Instrum. Meth. B **285**, 65 (2012)
30. Brewer, L.R., et al.: Phys. Rev. A **38**, 859 (1988)
31. Bollinger, J.J., et al.: Phys. Rev. A **48**(1), 525 (1993)
32. Bharadia, S., Vogel, M., Segal, D.M., Thompson, R.C.: Appl. Phys. B **107**, 1105 (2012)
33. Dubin, D.H.E., O'Neil, T.M.: Rev. Mod. Phys. **71**(1), 87 (1999)
34. Vogel, M., Winters, D.F.A., Segal, D.M., Thompson, R.C.: Rev. Sci. Instr. **76**, 103102 (2005)
35. Marshall, A.G., Hendrickson, C.L., Jackson, G.S.: Mass Spectrom. Rev. **17**, 1 (1998)
36. Keldysh, L.: Sov. Phys. JETP **20**, 1307 (1965)
37. Bélanger, P.A.: Opt. Lett. **16**, 196 (1991)
38. Ammosov, M.V., Delone, V.K.: Sov. Phys. JETP **64**, 1191 (1986)

71 Springer

Hyperfine Interact (2015) 236:73–77
DOI 10.1007/s10751-015-1177-1

Rate amplification of the two photon emission from para-hydrogen toward the neutrino mass measurement

Takahiko Masuda[1] · Hideaki Hara[1] · Yuki Miyamoto[1] · Susumu Kuma[1,5] ·
Itsuo Nakano[1] · Chiaki Ohae[2,6] · Noboru Sasao[1] · Minoru Tanaka[4] ·
Satoshi Uetake[3] · Akihiro Yoshimi[1] · Koji Yoshimura[1] ·
Motohiko Yoshimura[3]

Published online: 23 April 2015
© Springer International Publishing Switzerland 2015

Abstract We recently reported an experiment which focused on demonstrating the macro-coherent amplification mechanism. This mechanism, which was proposed for neutrino mass measurements, indicates that a multi-particle emission rate should be amplified by coherence in a suitable medium. Using a para-hydrogen molecule gas target and the adiabatic Raman excitation method, we observed that the two photon emission rate was amplified by a factor of more than 10^{15} from the spontaneous emission rate. This paper briefly summarizes the previous experimental result and presents the current status and the future prospect.

Keywords Neutrino · Raman excitation · Para-hydrogen

Proceedings of the 6th International Conference on Trapped Charged Particles and Fundamental Physics (TCP 2014), Takamatsu, Japan, 1–5 December 2014.

✉ Takahiko Masuda
masuda@okayama-u.ac.jp

[1] Research Core for Extreme Quantum World, Okayama University, Okayama 700-8530, Japan

[2] Graduate School of Natural Science and Technology, Okayama University, Okayama 700-8530, Japan

[3] Research Center of Quantum Universe, Okayama University, Okayama 700-8530, Japan

[4] Department of Physics, Osaka University, Toyonaka, Osaka 560-0043, Japan

[5] Present address: Atomic, Molecular and Optical Physics Laboratory, RIKEN, Wako, Saitama 351-0198, Japan

[6] Present address: Department of Engineering Science, University of Electro-Communications, Chofu, Tokyo 182-8585, Japan

1 Introduction

As Higgs particle was observed [1, 2], all 17 elementary particles in the Standard Model of elementary particle physics have been found. One of the basic parameters, masses of the particles, also have been determined. But the absolute masses of neutrinos are not determined yet because of the smallness of their masses and the weakness of their interaction. In addition to the fact that unknown parameters of particles should be determined, the mass of neutrino possibly relates the matter-antimatter asymmetry in our universe [3], which is one of the most interesting objectives in the particle physics.

Many experiments aiming to determine the neutrino mass are going on in the world. The current neutrinoless double beta decay experiments have sensitivity of $\sim 0.2 - 0.4$ eV if the neutrino is Majorana particle [4–6]. The direct mass measurement using tritium beta decay has set an upper limit of $m(\nu_e) < 2.05$ eV [7] and the next experiment which is being constructed will have sensitivity of 0.2 eV [8]. The cosmological surveys, on the other hand, have set an upper limit on the total neutrino mass of 0.28 eV [9].

Our group has proposed a new method of the neutrino mass measurement, which uses atom or molecule targets and a rate amplification mechanism by a macroscopic coherence [10]. The amplification by coherence, which we call macro-coherent amplification, is expected in a process of emitting plural particles. We recently reported an experiment which focused on demonstrating this amplification mechanism in the case of two photon emission process using para-hydrogen (p-H$_2$) gas target. The detail of this experiment and numerical simulation result can be found in [11]. This paper reports the summary of the previous experiment, a current status of the next experiment, and a future prospect.

2 Experiment

The brief schematic of the experimental setup is shown in Fig. 1. A p-H$_2$ gas at a temperature of 78 K and a pressure of 60 kPa was used as a target. It was filled in a copper cylinder 150-mm-long and 20 mm in diameter. Two laser pulses were injected into the p-H$_2$ target simultaneously. The properties of the lasers are summarized in Table 1. The output pulses from the target were measured by a MCT (Hg-Cd-Te) detector via a monochromator.

The relevant states of this experiment are the ground state ($Xv = 0$; 0 eV) and the electronically ground vibrationally excited state ($Xv = 1$; 0.5159 eV). Two laser pulses generated initial coherence between these states via the adiabatic Raman process [12]. We observed the anti-Stokes sidebands up to eighth order and the Stokes sidebands up to fourth order in this condition.

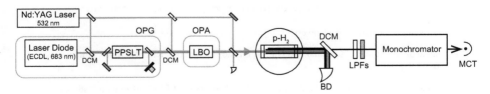

Fig. 1 Schematic of the experimental setup. PPSLT and LBO are nonlinear optical crystals used for the optical parametric generation (OPG) and the optical parametric amplification (OPA), respectively. ECDL: external cavity diode laser; DCM: dichroic mirror; BD: Beam dumper; LPFs: long-pass filters; MCT: Hg-Cd-Te mid-infrared detector

Table 1 Properties of two laser pulses

Wavelength	532.216 nm		683.610 nm
Pulse Energy in the Target	4.3 mJ		4.3 mJ
Beam Radius in the Target	0.12 mm		0.15 mm
Line Width	< 100 MHz		97 MHz
Pulse Duration	8 ns		6 ns
Polarization		horizontal	
Repetition Rate		10 Hz	

The energy of the fourth Stokes sideband (4662 nm ~ 0.266 eV) is lower than the transition energy, and thus it can stimulate the two photon emission from the excited state. We clearly observed two photon emission peaks in the output pulses from p-H_2 (Fig. 2). These spectra show the fourth Stokes signal at 4662 nm and its two-photon partner at 4959 nm.

The enhancement factor of the emission rate from the spontaneous emission was calculated based on the number of observed photons in 4959 nm peak and the number of excitable molecules irradiated with the laser pulses with corrections for the experimental acceptance (Table 2). The resultant enhancement factor of 10^{15} can be understood in the presence of macro-coherence.

3 Plan for experiment with the external trigger

In the experiment described in Section 2, we used the fourth Stokes light as a trigger laser to stimulate the two photon emission. Now we are developing the next experimental setup employing an external trigger laser. Figure 3 shows the schematic of the next experiment. There is a new 4.58 μm MIR pulse laser to stimulate the two photon emission. Using the

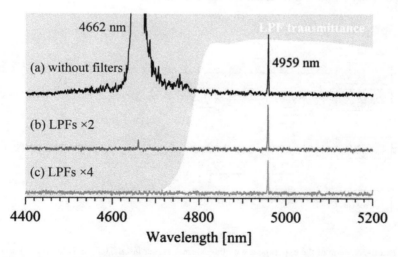

Fig. 2 Observed spectra of output light from p-H_2 target; **a** without the long-pass filter (LPF), **b** with two LPFs, and **c** with four LPFs. The *white portion* excluded by the *gray hatching* shows the LPF transmittance; it is ~ 0.85 at 4.96 μm [11]

Springer

Table 2 Parameters for the calculation of the enhancement factor [11]

Factor	Value
Spontaneous Decay Rate	3.2×10^{-11} s^{-1}
Energy Bandwidth of the Monochromator	$4.9 \times 10^{-3} \times 4959$ nm
Measurement Time	80 ns
Detector Solid Angle Fraction ($\Delta\Omega/4\pi$)	1.2×10^{-4}
Maximum Number of Excited Molecules	1.5×10^{16}
Calculated Spontaneous Emission (4959 nm)	$< 1.6 \times 10^{-8}$ photons/pulse
Observed Emission (4959 nm)	$> 4.4 \times 10^{7}$ photons/pulse

external trigger laser, we can control the coherence generation (i.e. excitation laser parameters) and the stimulating trigger condition (i.e. trigger laser parameters) independently. To understand this amplification mechanism more quantitatively and enlarge the enhancement factor, we will perform the further experiment and construct a realistic simulation model.

4 Summary

The coherent amplification mechanism for the plural particle emission is important for the future neutrino mass measurement. As easier case of the plural particle emission process, we used two photon emission from a para-hydrogen gas target to demonstrate the amplification mechanism. We have observed the amplification factor of more than 10^{15} from the spontaneous emission rate. We are now preparing the next experiment using the external trigger laser to study more detail of the mechanism toward the neutrino mass measurement.

One of the authors (T.M.) thanks the organizers of the TCP2014 conference for providing the opportunity for presenting about our experiment.

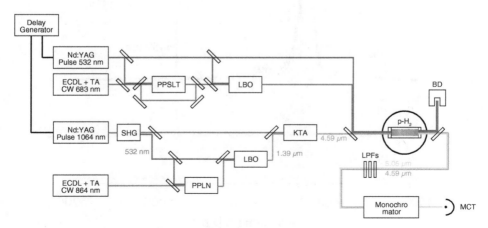

Fig. 3 Brief schematic of the experiment with the external trigger laser. 5.05 μm is the two-photon partner of the external trigger of 4.58 μm. PPSLT, LBO, PPLN, and KTA are nonlinear optical crystals. Delay generator adjusts the mutual timing between excitation two lasers of 532 nm and 683 nm and the trigger laser of 4.58 μm. SHG: second harmonic generator; TA: tapered amplifier

5 Compliance with ethical standards

This research was funded by Grant-in-Aid for Scientific Research on Innovative Areas "Extreme quantum world opened up by atoms" (21104002), Grant-in-Aid for Scientific Research A (21244032), Grant-in-Aid for Scientific Research C (25400257), Grant-in-Aid for Challenging Exploratory Research (24654132), Grant-in-Aid for Young Scientists B (25820144) and Grant-in-Aid for Research Activity start-up (26887026) from the Ministry of Education, Culture, Sports, Science, and Technology. The authors have no conflict of interest.

References

1. ATLAS Collaboration: Observation of a new particle in the search for the Standard Model Higgs boson with the ATLAS detector at the LHC. Phys. Lett. B **716**, 1–29 (2012). doi:10.1016/j.physletb.2012.08.020
2. CMS Collaboration: Observation of a new boson at a mass of 125 GeV with the CMS experiment at the LHC. Phys. Lett. B **716**, 30–61 (2012). doi:10.1016/j.physletb.2012.08.021
3. Davidson, S., Ibarra, A.: Leptogenesis and low-energy phases. J. Phys. G: Nucl. Part. Phys **29**, 1881–1883 (2003). doi:10.1088/0954-3899/29/8/366
4. Agostini, M., et al., (GERDA Collaboration): Results on neutrinoless double-β decay of ^{76}Ge from Phase I of the GERDA experiment. Phys. Rev. Lett **111**, 122503 (2013). doi:10.1103/PhysRevLett.111.122503
5. Gando, A., et al., (KamLAND-Zen Collaboration): Limit on neutrinoless $\beta\beta$ decay of ^{136}Xe from the first phase of KamLAND-Zen and comparison with the positive claim in ^{76}Ge. Phys. Rev. Lett. **110**, 062502 (2013). doi:10.1103/PhysRevLett.110.062502
6. The EXO-200 Collaboration: Search for Majorana neutrinos with the first two years of EXO-200 data. Nature **510**, 229–234 (2014). doi:10.1038/nature13432
7. Aseev, V.N., et al.: Upper limit on the electron antineutrino mass from the Troitsk experiment. Phys. Rev. D **112003**, 84 (2011). doi:10.1103/PhysRevD.84.112003
8. Bonn, J., et al.: The KATRIN sensitivity to the neutrino mass and to right-handed currents in beta decay. Phys. Lett. B **703**, 310–312 (2011). doi:10.1016/j.physletb.2011.08.005
9. Thomas, S.A., Abdalla, F.B., Lahav, O.: Upper Bound of 0.28 eV on Neutrino Masses from the Largest Photometric Redshift Survey. Phys. Rev. Lett **031301**, 105 (2010). doi:10.1103/PhysRevLett.105.031301
10. Fukumi, A., et al.: Neutrino spectroscopy with atoms and molecules. Prog. Theor. Exp. Phys. **2012**, 04D002 (2012). doi:10.1093/ptep/pts066
11. Miyamoto, Y., et al.: Observation of coherent two-photon emission from the first vibrationally excited state of hydrogen molecules. Prog. Theor. Exp. Phys. **2014**, 113C01 (2014). doi:10.1093/ptep/ptu094
12. Kien, F.L., et al.: Subfemtosecond pulse generation with molecular coherence control in stimulated Raman scattering. Phys. Rev. A **60**, 1562–1571 (1999). doi:10.1103/PhysRevA.60.1562

Hyperfine Interact (2015) 236:79–86
DOI 10.1007/s10751-015-1166-4

Highly charged ions for atomic clocks and search for variation of the fine structure constant

V. A. Dzuba[1] · V. V. Flambaum[1]

Published online: 24 March 2015

Abstract We review a number of highly charged ions which have optical transitions suitable for building extremely accurate atomic clocks. This includes ions from Hf^{12+} to U^{34+}, which have the $4f^{12}$ configuration of valence electrons, the Ir^{17+} ion, which has a hole in almost filled $4f$ subshell, the Ho^{14+}, Cf^{15+}, Es^{17+} and Es^{16+} ions. Clock transitions in most of these ions are sensitive to variation of the fine structure constant, α ($\alpha = e^2/\hbar c$). E.g., californium and einsteinium ions have largest known sensitivity to α-variation while holmium ion looks as the most suitable ion for experimental study. We study the spectra of the ions and their features relevant to the use as frequency standards.

Keywords Variation of alpha · Optical clock · Highly charged ion

PACS 06.30.Ft · 06.20.Jr · 31.15.A · 32.30.Jc

1 Introduction

Highly charged ions (HCI) can be used for building a new generation of very accurate optical clocks [1–3]. This may have many technical applications but also clock transitions in HCI can be used to study fundamental problems of modern physics such as variation of fundamental constants [4], local Lorentz invariance violation [5, 6], search for dark matter [7], etc. Having extremely high accuracy of the clocks is crucial for these studies. Clocks

Proceedings of the 6th International Conference on Trapped Charged Particles and Fundamental Physics (TCP 2014), Takamatsu, Japan,1–5 December 2014.

✉ V. A. Dzuba
V.Dzuba@unsw.edu.au

[1]   School of Physics, University of New South Wales, Sydney 2052, Australia

Table 1 Electric quadrupole (E2) clock transition

Valence configuration	Ultra-relativistic limit	Transition (ground state - clock state)	Q
p^2	$p_{1/2}^2$	$^3P_0-{^1D_2}$ or $^3P_0-{^1S_0}$	$< 10^{19}$
p^4	$p_{3/2}^2$	$^3P_0-{^1S_0}$	$< 10^{19}$
d^2	$d_{3/2}^2$	$^3F_0-{^3P_0}$	$< 10^{19}$
d^8	$d_{5/2}^2$	$^3F_4-{^3P_2}$	$\sim 10^{19}$
f^2	$f_{5/2}^2$	$^3H_4-{^3F_2}$	$\sim 10^{19}$
f^{12}	$f_{7/2}^2$	$^3H_6-{^3F_4}$	$\sim 10^{20}$

quality factor can be defined as a ratio of clock frequency to the value of its perturbations

$$Q = \omega/\delta\omega. \tag{1}$$

Current microwave cesium clock, which serves as definition of metric second, have $Q \sim 10^{16}$ [8], best optical clocks approach $Q \sim 10^{18}$ [9–11] mostly due to larger frequency. Further progress can be achieved by using optical transitions in HCI [1, 2]. HCI are less sensitive to perturbations due to their compact size. Therefore, quality factor $Q = \omega/\delta\omega$ can be larger than in neutral atoms due to smaller $\delta\omega$. In this paper we review some recent proposals for very accurate atomic clocks based on HCI and their use for the search of time variation of the fine structure constant.

2 The $4f^{12}$ ions.

Optical transitions in HCI can be easily found between states of the same configuration. All such transitions must be even-parity transitions. Even electromagnetic transitions include magnetic dipole (M1), electric quadrupole (E2), and higher-order transitions. Magnetic dipole is usually too strong to be used in clock transitions. Accuracy of the clocks would be limited by natural width of the line. On the other hand, higher-order transitions are too weak and not very convenient to work with. The best candidates seem to be electric quadrupole transitions (E2). The E2 transitions suitable for the use as clock transitions can be found in configurations consisting of two identical electrons or holes in an almost filled subshell (see Table 1).

The best candidates seems to be the ions with the $4f^{12}$ configuration of valence electrons in the ground state [2]. All HCI from Hf^{12+} to U^{34+} fell in this category. Typical energy diagram is presented on Fig. 1 for Os^{18+}.

There is an electric quadrupole clock transition between the $^3H_6$ and $^3F_4$ states as well as magnetic dipole transitions from both ground and clock states, which can be used for cooling and/or detection.

Both states of the clock transition have non-zero quadrupole moments, which make them sensitive to gradients of electric field. This may affect the accuracy of the clock if not addressed. There are different ways to deal with the problem. One is by using the hyperfine structure (hfs). If we take an ion with non-zero nuclear moment, there is a good chance to find hyperfine states with almost identical quadrupole energy shift in upper and lower states [1]. Then the shifts cancel each other in the frequency of the transition (see diagram for $^{209}Bi^{25+}$ on Fig. 2).

Fig. 1 Low-lying energy levels
of Os^{18+}

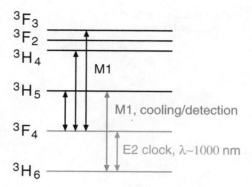

This way of dealing with the quadrupole shift has a shortcoming, relatively high sensitivity to the second-order Zeemen shift. This is due to enhancement of the shift by small energy denominators, which are the hfs intervals.

Another way of dealing with the problem is by choosing isotopes with zero nuclear spin (and no hyperfine structure) [2]. Here one can make a combination of frequencies between different Zeeman states, which is not sensitive to the electric qudrupole shift, see Fig. 3. One has to know the ratio of quadrupole moments of the two clock states to find the right combination [2]. Magnetic field is to be used to separate Zeeman states. First order Zeeman shift cancells out in the transition with the same value but opposite sign of the projection M of the total angular momentum J. Second-order Zeeman shift is small.

3 Search for variation of the fine structure constant.

The possibility of the fundamental constants to vary is suggested by theories unifying gravity with other fundamental interactions (see, e.g. [12]). The study of quasar absorption spectra indicates that the fine structure constant α ($\alpha = e^2/\hbar c$) may vary in space or time. In this study a quasar serves as a powerful source of light in wide range of spectra. Part of this light is absorbed on the way to Earth by a gas cloud brining to Earth informations about atomic spectra billions of years ago. When all the differences in the quasar absorption spectra and atomic spectra obtained in the laboratory are significantly reduced by adjusting the value of just one parameter, the fine structure constant, this is considered as evidence of slightly different value of α at distant past (or at long distance). The analysis of huge amount of data coming from two telescopes, Kerk telescope in Hawaii and VLT in Chile, reveals that α may vary on astronomical scale along a certain direction in space forming the so called alpha-dipole [13]

$$\frac{\Delta\alpha}{\alpha} = (1.1 \pm 0.2) \times 10^{-6}\,\mathrm{Gly}^{-1} \cdot r\cos\theta, \qquad (2)$$

where r and θ are the coordinate of a point in space in the framework of alpha-dipole.

Fig. 2 Clock states of Bi^{25+}, including hiperfine structure

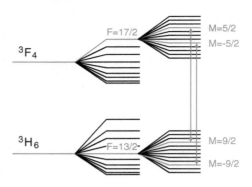

Earth movements in the framework of alpha-dipole leads to time variation of α in laboratory [14]

$$\frac{1}{\alpha}\frac{\partial\alpha}{\partial t} = \left[1.35 \times 10^{-18}\cos\psi + \right.$$
$$\left. 1.4 \times 10^{-20}\cos\omega t\right]\mathrm{y}^{-1} \approx 10^{-19}\mathrm{y}^{-1}. \tag{3}$$

Here ψ is the angle between alpha-dipole and direction of Sun movement, ($\cos\psi \approx 0.07$); second oscillating term in (3) is due to Earth movement around Sun.

So small rate of change suggests that the most precise atomic clocks are probably the only adequate tools for detecting it. Indeed, the best current limit on the time-variation of α comes from comparing Al$^+$ and Hg$^+$ optical clocks over long period of time [15],

$$\frac{1}{\alpha}\frac{\partial\alpha}{\partial t} = (-1.6 \pm 2.3) \times 10^{-17}\mathrm{y}^{-1}. \tag{4}$$

The accuracy of the most precise optical clocks approaches the level of 10^{-18} [9–11]. However, it does not immediately lead to similar sensitivity to variation of α. The relative change in clock frequency due to variation of α can be expressed as

$$\frac{1}{\omega}\frac{\partial\omega}{\partial t} = K\frac{1}{\alpha}\frac{\partial\alpha}{\partial t}, \tag{5}$$

where K is electron structure factor which comes from atomic calculations. It turns out that for best optical clocks $K < 1$ ($K = 0.31$ for Yb, $K = 0.062$ for Sr [16], and $K = -0.3$ for the f^{12} ions considered above). Therefore, we need to search for systems, which have all features of best atomic clocks but also sensitive to variation of α, i.e., $K \gg 1$.

It is convenient to present dependence of atomic frequencies on α in a form

$$\omega = \omega_0 + q\left[(\alpha/\alpha_0)^2 - 1\right], \tag{6}$$

where q is the electron structure factor describing relativistic frequency shift. Comparing (6) with (5) leads to $K = 2q/\omega_0$. There are two ways of searching for large K, look for large frequency shift q, or look for small frequency ω_0. The best known example of the second kind is dysprosium atom, which has a pair of degenerate states of opposite parity for which $K \sim 10^8$ [17]. This pair of states was used indeed in the search for time-variation of the fine structure constant α [18]. The result

$$\frac{1}{\alpha}\frac{\partial\alpha}{\partial t} = (-5.8 \pm 6.9) \times 10^{-17}\mathrm{y}^{-1}. \tag{7}$$

Fig. 3 Clock states of Os^{18+}, including Zeeman splitting

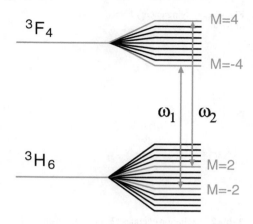

puts slightly weaker limit on time variation of α than the Al^+/Hg^+ clocks (4). This is in spite of huge relative enhancement for Dy and almost no enhancement for Al^+/Hg^+. The reason is that the degenerate states of dysprosium lack the features of an atomic clock transition, e.g. one of the states is pretty short-living. Therefore, what is gained on the enhancement is lost on the accuracy of frequency measurements.

In this work we study another possibility. Keep ω_0 in optical region to take full advantage of extremely accurate optical clocks, and find systems with large relativistic energy shift q. To see where such systems can be found, it is instructive to use an analytical estimate for relativistic energy shift [19]

$$\Delta E \approx \frac{E}{\nu} (Z\alpha)^2 \left(\frac{1}{j + 1/2} - C \right), \qquad (8)$$

where ν is the effective principal quantum number ($E = -1/2\nu^2$), Z is nuclear charge, j is total angular electron momentum, and C is semi-empirical factor to simulate many-body effects in many-electron atoms ($C \sim 0.6$). Relativistic frequency shift for a transition between states a and b is given by $q = \Delta E_a - \Delta E_b$. One can see from (8) that large frequency shift q can be found in heavy (large Z) highly charged ions, where $E \sim (Z_i + 1)^2$ (Z_i is ionization degree), in transitions, which correspond to $s - f$ or $p - f$ single-electron transitions (largest change of j) [4]. One problem here is that such transitions are usually not optical since energy intervals grow very fast with ionization degree Z_i ($\Delta E \sim (Z_i + 1)^2$). The solution comes from level crossing [4, 20, 21]. Since state ordering in neutral atoms and hydrogen-like ions is different, there are must be change in ordering of s and f or p and f states at some intermediate ionization degree. In the vicinity of level crossing the frequencies of corresponding transitions are likely to be in optical region [20].

A number of optical transitions in HCI sensitive to variation of the fine structure constant were considered in Ref. [4, 21, 23–25]. However, most of these transitions lack some features of atomic clock transitions, which limit the accuracy of frequency measurements. Important questions of ions trapping and cooling, preparation and detection of states, etc. were also not discussed. All these questions were first addressed in recent paper [26]. In particular, the criteria for good clock system, sensitive to variation of α, were formulated. The main points include: (a) In terms of single-electron transitions, the clock transition is a $s - f$ or $p - f$ transition. This makes it sensitive to the variation of α. (b) the frequency of the transition is in optical region (5000 cm^{-1} < $\hbar\omega$ < 43000 cm^{-1}, or 230 nm < λ < 2000 nm).

Fig. 4 Low-lying energy levels of Ir^{17+}

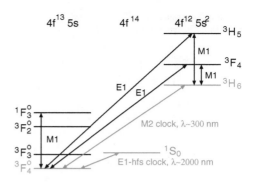

Table 2 Excitation energies (E, cm^{-1}), sensitivity factors (q, cm^{-1}), and enhancement factors ($K = 2q/E$) for the clock states of Ir^{17+} ion

N	Conf.	Term	E	q	K
1	$4f^{13}5s$	$^2F^o_4$	0	0	0
2	$4f^{14}$	$^1S_0$	5000	370000	150
3	$4f^{12}5s^2$	$^3H_6$	30000	-390000	-26

(c) This is a transition between ground and a metastable state with lifetime between 100 and 10^4 seconds. (d) There are other relatively strong optical transitions with transition rate $\gtrsim 10^3$ s^{-1}. (e) The transition is not sensitive to perturbations such as gradients of electric field, Zeeman shift, black-body radiation shift, etc.

A very promising systems is the Ho^{14+} ion [26]. It has following features: (a) Clock transition between the $4f^65s$ $^8F_{1/2}$ and $4f^55s^2$ $^6H^o_{5/2}$ states is sensitive to variation of alpha (the $4f - 5s$ transition). (b) It is optical transition ($\lambda \approx 400$ nm). (c) It is a narrow transition from ground state to a metastable state. (d) There are electric dipole (E1) and magnetic dipole (M1) transitions from both ground and clock states. (e) The clock transition can be made insensitive to gradients of electric field, which are coupled to atomic quadrupole moment. This can be done by proper choice of the values of the total angular momentum F (including nuclear spin I, $F = J + I$) and its projection M. Quadrupole shift disappears for $F = 3$ and $M = 2$ since it is proportional to $3M^2 - F(F + 1)$. Experimental work with the Ho^{14+} ions is in progress at RIKEN [27].

There must be at least two clocks to register variation of alpha since such variation can only be unambiguously detected in a dimensionless ratio of two frequencies. A good option is to have one clock transition, which is sensitive to variation of α and another, which is not. Note that cesium clock is not good enough if we want relative accuracy of the order 10^{-18}. A good option might come from the use of HCI with the $4f^{12}$ configuration of valence electrons considered in Section 2, e.g. the Os^{18+} ion. Comparing the Ho^{14+} and Os^{18+} clocks provides high sensitivity to variation of alpha ($K \approx -18$).

Another option is to use the Ir^{17+} ions. Clock transitions in these ions involve hole states in the $4f$ subshell which leads to extra enhancement of the sensitivity of clock frequencies to variation of the fine structure constant [21]. A diagram for few lowest states of Ir^{17+} is presented on Fig. 4. The energies and sensitivity coefficients for clock states of Ir^{17+} ions

Table 3 Long-living isotopes of Cf and Es and clock transitions in Cf^{15+}, Es^{17+} and Es^{16+}

Isotope			Clock transition				$\hbar\omega$	q	K
Ion	Lifetime	I	Ground state		Clock state		cm^{-1}	cm^{-1}	
$^{249}Cf^{15+}$	351 y	9/2	$5f6p^2$	$^2F^o_{5/2}$	$5f^26p$	$^2H^o_{9/2}$	13303	380000	57
$^{252}Es^{17+}$	1.29 y	5	$5f^2$	$^3H_4$	$5f6p$	$^3F_2$	7017	-46600	-13
$^{253}Es^{16+}$	20 d	7/2	$5f^26p$	$^4I^o_{9/2}$	$5f6p^2$	$^2F^o_{5/2}$	7475	-184000	-49
$^{255}Es^{16+}$	40 d	7/2	$5f^26p$	$^4I^o_{9/2}$	$5f6p^2$	$^2F^o_{5/2}$	7475	-184000	-49

are presented in Table 2. Note, that the ion has two clock transitions which have different sensitivity to variation of α. If the ratio of two frequencies is measured, the combined sensitivity is very large

$$\frac{\partial}{\partial t} \ln \frac{\omega_2}{\omega_1} = (K_2 - K_1) \frac{1}{\alpha} \frac{\partial \alpha}{\partial t} = -176 \frac{1}{\alpha} \frac{\partial \alpha}{\partial t}. \tag{9}$$

Experimental work with Ir^{17+} ions is in progress at Max Planck Institute [28].

It has been mentioned above that we are looking for systems with large sensitivity coefficients q and clock frequency ω being in optical region so that the enhancement factor K ($K = 2q/\omega$) is large due to large q rather than small ω. The largest sensitivity factors found so far are in the Cf^{17+} and Cf^{16+} ions [29]. However, both these ions are not very convenient for the use as atomic clocks. The Cf^{17+} ion has only one optical transition other than clock transition. This is a magnetic dipole transition between ground $5f_{5/2}$ state and excited $5f_{7/2}$ state. It is weak due to small frequency of the transition. This makes it difficult working with the ion. The Cf^{16+} ion has only one metastable excited state which can serve as a clock state. The transition between ground $5f6p$ $^1F_3$ and metastable $6p^2$ $^1S_0$ states is the magnetic octupole (M3) transition. It is so weak that its use as clock transition is very problematic.

It turns out that good clock transitions exist in the Cf^{15+} ion as well as in the Es^{16+} and Es^{17+} ions. The main parameters of the ions and corresponding clock transitions are summarized in Table 3. Note that both elements have only unstable isotopes. However, many isotopes have very long lifetime (see, e.g. Table 3). The choice of isotopes in Table 3 is dictated by two considerations. First, it is a long-living isotope. Second, nuclear spin I has such a value that it is always possible to have $F = 3$, $M = 2$ for both states in the clock transition. Here $F = I + J$ is the total angular momentum of the ion and M is its projection. States with $F = 3$ and $M = 2$ have zero quadrupole moment and are not sensitive to gradients of electric field.

Acknowledgements The authors are grateful to H. Katori and M. Wada for useful discussions. The work was supported by the Australian Research Council.

Conflict of interests There is no conflict of interest. No human participants or animals were used. No copyright material was used.

References

1. Derevianko, A., Dzuba, V.A., Flambaum, V.V.: Phys. Rev. Lett. **109**, 180801 (2012)
2. Dzuba, V.A., Derevianko, A., Flambaum, V.V.: Phys. Rev. A **86**, 054501 (2012)

3. Dzuba, V.A., Derevianko, A., Flambaum, V.V.: Phys. Rev. A **87**, 029906(E) (2013)
4. Berengut, J.C., Dzuba, V.A., Flambaum, V.V.: Phys. Rev. Lett. **105**, 120801 (2010)
5. Hohensee, M.A., Leefer, N., Budker, D., Harabati, C., Dzuba, V.A., Flambaum, V.V.: Phys. Rev. Lett. **111**, 050401 (2013)
6. Pruttivarasin, T., Ramm, M., Porsev, S.G., Tupitsyn, I.I., Safronova, M.S., et al.: Nature **517**, 592 (2015)
7. Derevianko, A., Pospelov, M.: Nat. Phys. **10**, 933 (2014)
8. http://www.nist.gov/pml/div688/grp50/primary-frequency-standards.cfm
9. Hinkley, N., Sherman, J.A., Phillips, N.B., Schioppo, M., Lemke, N.D., Beloy, K., Pizzocaro, M., Oates, C.W., Ludlow, A.D.: Science **341**, 1215 (2013)
10. Bloom, B.J., Nicholson, T.L., Williams, J.R., Campbell, S.L., Bishof, M., Zhang, X., Zhang, W., Bromley, S.L., Ye, J., et al.: Nature **506**, 71 (2014)
11. Ushijima, I., Takamoto, M., Das, M., Ohkubo, T., Katori, H.: Nat. Photonics **9**, 185 (2015)
12. Uzan, J.-P.: Rev. Mod. Phys. **75**, 403 (2003)
13. Webb, J.K., King, J.A., Murphy, M.T., Flambaum, V.V., Carswell, R.F., Bainbridge, M.B.: Phys. Rev. Lett. **107**, 191101 (2011)
14. Berengut, J., Flambaum, V.V.: Europ. Phys. Lett. **97**, 20006 (2012)
15. Rosenband, T., et al.: Science **319**, 1808 (2008)
16. Angstmann, E.J., Dzuba, V.A., Flambaum, V.V.: Phys. Rev. A **70**, 014102 (2004)
17. Dzuba, V.A., Flambaum, V.V.: Phys. Rev. A **77**, 012515 (2008)
18. Leefer, N., Weber, C.T.M., Cingöz, A., Torgerson, J.R., Budker, D.: Phys. Rev. Lett. **111**, 060801 (2013)
19. Dzuba, V.A., Flambaum, V.V., Webb, J.K.: Phys. Rev. A **59**, 230 (1999)
20. Berengut, J.C., Dzuba, V.A., Flambaum, V.V., Ong, A.: Phys. Rev. A **86**, 022517 (2012)
21. Berengut, J.C., Dzuba, V.A., Flambaum, V.V., Ong, A.: Phys. Rev. Lett. **106**, 210802 (2011)
22. Dzuba, V.A., Derevianko, A., Flambaum, V.V.: Phys. Rev. A **86**, 054502 (2012)
23. Safronova, M.S., Dzuba, V.A., Flambaum, V.V., Safronova, U.I., Porsev, S.G., Kozlov, M.G.: Phys. Rev. Lett. **113**, 030801 (2014)
24. Safronova, M.S., Dzuba, V.A., Flambaum, V.V., Safronova, U.I., Porsev, S.G., Kozlov, M.G.: Phys. Rev. A **90**, 042513 (2014)
25. Safronova, M.S., Dzuba, V.A., Flambaum, V.V., Safronova, U.I., Porsev, S.G., Kozlov, M.G.: Phys. Rev. A **90**, 052509 (2014)
26. Dzuba, V.A., Flambaum, V.V., Katori, H.: Phys. Rev. A **91**, 022119 (2015)
27. Katori, H., Nakamura, N., Okada, K.: private communication
28. Windberger, A., Versolato, O.O., Bekker, H., Oreshkina, N.S., Berengut, J.C., Bock, V., Borschevsky, A., Dzuba, V.A., Harman, Z., Kaul, S., Safronova, U.I., Flambaum, V.V., Keitel, C.H., Schmidt, P.O., Ullrich, J., Crespo Lopez-Urrutia, J.R.: Identification of optical transitions in complex highly charfed ions for applications in metrology and tests of fundamental constants. to be published
29. Berengut, J.C., Dzuba, V.A., Flambaum, V.V., Ong, A.: Phys. Rev. Lett. **109**, 070802 (2012)

Hyperfine Interact (2015) 236:87–94
DOI 10.1007/s10751-015-1188-y

Characterization of ion Coulomb crystals for fundamental sciences

**Kunihiro Okada[1] · Masanari Ichikawa[2] ·
Michiharu Wada[1]**

Published online: 13 May 2015
© Springer International Publishing Switzerland 2015

Abstract We performed classical molecular dynamics (MD) simulations in order to search
the conditions for efficient sympathetic cooling of highly charged ions (HCIs) in a linear
Paul trap. Small two-component ion Coulomb crystals consisting of laser-cooled ions and
HCIs were characterized by the results of the MD simulations. We found that the spatial distribution is determined by not only the charge-to-mass ratio but also the space charge effect.
Moreover, the simulation results suggest that the temperature of HCIs do not necessarily
decrease with increasing the number of laser-cooled ions in the cases of linear ion crystals.
We also determined the cooling limit of sympathetically cooled ^{165}Ho^{14+} ions in small linear ion Coulomb crystals. The present results show that sub-milli-Kelvin temperatures of at
least 10 Ho^{14+} ions will be achieved by sympathetic cooling with a single laser-cooled Be$^+$.

Keywords Sympathetic cooling · Highly charged ion · Ion trap

1 Introduction

Sympathetically cooled molecular ions and highly charged ions (HCIs) are fascinating
research objects for studying fundamental sciences, such as cold/ultracold ion chemistry

Proceedings of the 6th International Conference on Trapped Charged Particles and Fundamental
Physics (TCP 2014), Takamatsu, Japan,1-5 December 2014

✉ Kunihiro Okada
okada-k@sophia.ac.jp

Michiharu Wada
mw@riken.go.jp

[1] Department of Physics, Sophia University, 7-1 Kioicho, Chiyoda, Tokyo 102-8554, Japan

[2] RIKEN Nishina Center for Accelerator-Based Science, 2-1 Hirosawa, Wako, Saitama 351-0198,
Japan

[1, 2] and the study of possible time variations of fundamental constants via precise optical spectroscopy [3, 4]. Recently, cold highly charged ions were considered to be good candidates for probing the time variation of the fine structure constant α [5–7]. In performing these studies, the sympathetic cooling by ion Coulomb crystals in a linear Paul trap is a promising method for generating cold molecular ions or cold highly charged ions. Since ultra-high accuracy measurements of optical transitions are indispensable for studying possible time variation of α, for sympathetically cooled HCIs the so-called Lamb-Dicke criterion should be satisfied in a linear Paul trap. Thus, for such measurements, a linear configuration of ion crystals is much more suitable than large two-component Coulomb crystals with shell structures.

Recently precise laser spectroscopy of Ar^{13+} and Ir^{17+} ions by the sympathetic cooling technique was proposed to test quantum electrodynamics and to probe possible time variation of α [8, 9], and then a cryogenic linear Paul trap connecting with the EBIT facility was developed [10]. Trapping and cooling of externally injected Ar^{13+} into the cryogenic linear Paul trap was reported very recently [11, 12]. However, the direct observation of laser-induced fluorescence from crystallized HCIs has not been performed yet.

In this work, we performed molecular dynamics (MD) simulations in order to search the conditions for efficient sympathetic cooling of HCIs in a linear Paul trap. The characterization of small two-component linear ion Coulomb crystals consisting of HCIs and laser-cooled ions (LCIs) was performed using the simulation results. First we discuss the spatial distribution of sympathetically crystallized HCIs in a linear Paul trap. In particular the ion distribution of very highly charged $^{165}Ho^{66+}$ ions is discussed. Then the sympathetic cooling efficiency of HCIs via laser-cooled Be^+ ions for small ion crystals is investigated in detail by changing the simulation conditions. Finally we investigate the cooling limit of sympathetically cooled HCIs, namely $^{165}Ho^{14+}$, which is a promising candidate for detecting possible time variation of the fine structure constant [13].

2 Molecular dynamics simulation

We performed molecular dynamics simulations to characterize two-component linear ion Coulomb crystals in a linear Paul trap. In the present simulations, all forces, i.e., static and time-dependent quadrupole electric fields, radiation-pressure forces, average recoil forces by spontaneous emissions, and ion-ion Coulomb repulsion forces are taken into account in Newton's equations of motion of all the trapped ions. The equations are numerically integrated by the fourth-order Runge-Kutta algorithm with a 2 ns time step. In order to reduce a computational time for obtaining the quasi-equilibrium states of Coulomb crystals, we introduced cold elastic collisions between laser-cooled ions (LCIs) and virtual very light atoms in the early stage of integration steps. The details of the method is described in [14]. After this process, only the radiation pressure forces are applied to LCIs. The sympathetic cooling of HCIs is achieved by only the Coulomb interactions with LCIs.

Since the positions of all ions at each integration step are recorded, the simulation image of crystallized ions like an observed laser-induced fluorescence image is produced by the density plot of existence probabilities of the ions as follows. First we divide an predetermined image area into many small cells and the content of each cell is set to 0. Then, at each integration step, we increment a cell counter by one if an ion exists in this cell. The spatial distribution of ions can be evaluated by the production image. We also obtain the

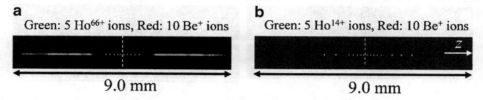

Fig. 1 Simulation images of two-component linear ion crystals consisting of 10 Be^+ and 5 Ho^{q+} ions: (a) $q = 66$, (b) $q = 14$. The vertical dotted line indicates the center of the trap along the z-axis. Simulation parameters: $f_{RF} = 10$ MHz, $V_{ac} = 30$ V, $V_z = 0.1$ V. The trapping parameters are $q(Ho^{66+}) = 0.123$, $q(Ho^{14+}) = 0.0262$, and $q(Be^+) = 0.0343$, respectively. The dimensions of the linear Paul trap are as follows: $r_0 = 2.18$ mm, $z_0 = 25$ mm, where r_0 and $2z_0$ are the closest distance from the trap axis to the rod electrode and the axial length of the ion trap, respectively. The geometrical factor (κ) along the trap axis was taken as 0.33. The intensity and the detuning of the cooling laser for calculations of radiation pressure forces were set to 3 mW/cm^2 and -10 MHz, respectively. The incidence angle of the cooling laser is 5 deg. with respect to the trap axis

micromotion energies of HCIs by averaging the kinetic energies of the trapped ions. The present simulations were performed by modifying the previously developed codes [14, 15].

3 Results and discussions

3.1 Spatial distribution of sympathetically cooled HCIs

In the case of singly charged ions, the ions with a higher charge-to-mass ratio (Q/m) are subjected to a stronger trapping force and consequently gather near the trap axis under the influence of the sympathetic cooling effect. However this is not the case for two-component Coulomb crystals including highly charged ions. Figure 1a shows a simulation image consisting of 10 Be^+ and 5 $^{165}Ho^{66+}$ ions at the quasi-equilibrium state. Although (Q/m) of Ho^{66+} is larger than that of Be^+, the HCIs are distributed in the outside of the LCIs. Since the Be^+ ions firstly occupy the positions around the minimum of the trapping potential by the cooling effect, the sympathetically cooled Ho^{66+} ions are pushed to the outside of the trap center by the space charge of the cold Be^+ ions.

It is to be noted that the size of the linear ion crystal is considerably larger than those of normal two-component linear ion crystals consisting of singly charged ions [14, 16]. This is also due to the strong Coulomb interactions between Ho^{66+} and Be^+ ions. Actually the axial size of a linear ion crystal becomes smaller as the charge state decreases, as shown in Fig. 1b. On the one hand, we need to increase the axial static voltage (V_z) applied along the trap axis in order to compress the axial distribution. However, the high V_z leads to breaking of the string shape of the ion crystal and then induces a high rf heating rate. In fact the Ho^{66+} ions are easily lost from the trap by increasing V_z from 0.1 to 1.0 V in the simulation conditions of Fig. 1a. It is also important to reduce asymmetric fields in the radial direction as much as possible. For the explanation of these points see also in Section 3.2.

Figure 2 shows the coolant ion-number dependence of the average kinetic energy of 6 Ho^{66+} ions. In these simulations, the average kinetic energy of the coolant Be^+ ions was maintained to be 23 mK by applying cold elastic collisions with virtual very light atoms. The reasons why this method works well to simulate the Coulomb crystals are described in the previous paper [14]. The temperatures indicated in the images in Fig. 2 show the average kinetic energies of 6 Ho^{66+} ions expressed by the unit of Kelvin. In the present simulation

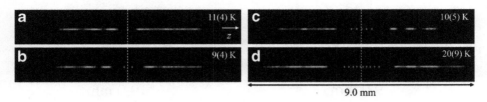

Fig. 2 Simulation images of two-component Coulomb crystals consisting of 6 Ho^{66+} ions (*green*) and the different numbers of Be$^+$ ions (*red*), where the average kinetic energy of the coolant ions is maintained to be 23 mK by cold elastic collisions with virtual very light atoms [14]. The vertical dotted line indicates the center of the trap along the z-axis. The other simulation parameters are the same as in Fig. 1

conditions, the average kinetic energy of the Ho^{66+} ions is almost same within the standard deviations in the case that the number of the Be$^+$ is up to 5. However, the average kinetic energy of Ho^{66+} increases as the number of Be$^+$ ions increases, as shown in Fig. 2d.

This phenomenon is possibly explained as in the following. The space charge of the coolant ions decreases the trapping potential depth but creates a barrier around the trap center. As the number of the coolant increases, the radial deviations of either coolant or HCIs become larger under larger axial stress from the extended string, leading to larger rf heating rates especially in light of the off-axis component of the cooling laser force. Even if the deviation from the trap axis is small, the large rf heating effect on the HCIs is expected. Thus, the excess number of the coolant ions possibly leads to low sympathetic cooling efficiency.

Note that it might be possible to take other ion configurations in small linear ion crystals consisting of Ho^{66+} and Be$^+$ ions as shown in Figs. 1 and 2. The configurations possibly depend on the initial conditions of the MD simulations [11, 16, 17]. Although the observed configurations are stable at least and are taking the local minimum of the energies under the present simulation conditions, it is not clear whether those are taking the minimum energy configurations or not. Moreover, it might be possible to take more stable configurations by collisions with residual gases [18] or by intensity fluctuations of lasers. The former effect may also change the charge state of HCIs. Since in the present simulations such the effects were not included, it needs further studies.

3.2 Cooling limit of sympathetically cooled Ho^{14+} with a single Be$^+$

It is interesting to know the cooling limit of Ho^{14+} by sympathetic cooling, since the Ho^{14+} is considered to be a good candidate to probe possible time variation of the fine structure constant via optical atomic clock [13]. Here we perform extensive MD simulations in order to search the conditions for efficient sympathetic cooling of Ho^{14+} in a linear Paul trap. As described in Section 3.1 the excess number of coolant ions leads to low sympathetic cooling efficiency. Therefore we first tested a single Be$^+$ ion as a coolant. Figure 3a shows plots of the average kinetic energies of Ho^{14+} and a single laser-cooled Be$^+$ as a function of time. The inset figure shows the position of each ion and the incident direction of the cooling laser. In this simulation, we applied the following calculation procedures.

In the period I, the cold elastic collisions between Be$^+$ and virtual very light atoms with thermal energy of 7 mK were introduced [14] and the sympathetic cooling of Ho^{14+} ions is achieved by the Coulomb interaction with Be$^+$. This approach drastically reduces calculation steps in order to achieve the quasi-equilibrium state of the two-component Coulomb crystal. This collision cooling method is useful to reproduce the realistic Coulomb crystals,

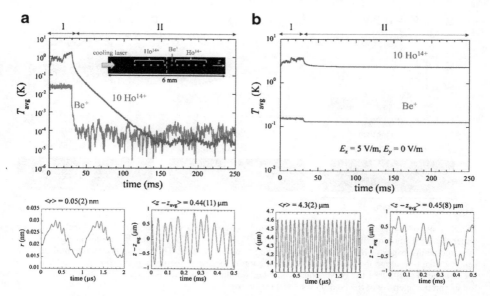

Fig. 3 Plots of average kinetic energies of Ho^{14+} and Be^+ as a function of time under the ideal trapping fields (a) and under applying the asymmetric static filed of $E_x = 5$ V/m along the x axis (b). The inset figure shows the position of each ion and the incident direction of the cooling laser. In the period I, the cold elastic collisions between Be^+ and virtual very light atoms with a thermal energy of 7 mK were introduced [14]. Then, in the period II, only the laser radiation pressure force was applied to Be^+. The lower figures show plots of the time variation of r and $z - z_{avg}$ of a certain Ho^{14+} ion, where z_{avg} is the average ion position in the z direction. We also show the average values of r and $z - z_{avg}$ at the quasi-equilibrium state of the linear ion crystal. The intensity and the detuning of the cooling laser were set to 1 mW/cm^2 and -10 MHz, respectively. The ion trap parameters: $f_{RF} = 10$ MHz, $V_{ac} = 80$ V, $V_z = 0.1$ V. The other simulation parameters are the same as in Fig. 1

where the cooling and heating effects are balancing each other due to occasional collisions with background gases [14]. However, it is not appropriate for the present purpose, i.e. searching for the sympathetic cooling limit of Ho^{14+}. Thus, in the period II, only the radiation pressure force was applied to Be^+.

We have successfully obtained the efficient cooling conditions of 10 Ho^{14} by a single laser-cooled Be^+ under the ideal trapping fields. The average kinetic energies of Ho^{14+} and Be^+ are 28(8) μK and 98(56) μK, respectively. The reason why the average kinetic energies appears to be lower than the Doppler cooling limit of Be^+,435 μK, is that the residual kinetic energy is distributed to the individual ions. In actual, the sum of the individual ion energy is $3.8(1.4) \times 10^2$ μK, which is consistent with the Doppler cooling limit of Be^+. The average radial position of Ho^{14+} is also obtained to be 0.05(2) nm. The position deviation of Ho^{14+} ions from the trap axis is supposed to be caused by the intentional cooling laser misalignment of 5 deg. It is noted that we can not obtain the actual time to achieve the cooling limit from the present simulations, since the purpose is the search for the cooling conditions and the cooling limit of Ho^{14+}.

In order to investigate the effect of imperfect radial electric fields caused by patch potentials or misalignment of the trap electrodes, we perform the MD simulation by applying a static electric field (E_x) in the x direction. Figure 3b is a plot of the average kinetic energy of 10 Ho^{14+} sympathetically cooled by a single Be^+. The simulation parameters are the same

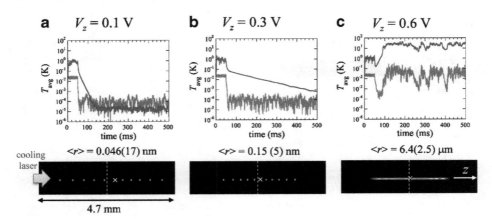

a $V_z = 0.1$ V **b** $V_z = 0.3$ V **c** $V_z = 0.6$ V

cooling laser

$\langle r \rangle = 0.046(17)$ nm $\langle r \rangle = 0.15\,(5)$ nm $\langle r \rangle = 6.4(2.5)\,\mu$m

4.7 mm

Fig. 4 Plots of average kinetic energies of 10 Ho^{14+} (*blue*) and a single Be$^+$ (*red*) as a function of time under the ideal trapping fields: (a) $V_z = 0.1$ V, (b) $V_z = 0.3$ V, (c) $V_z = 0.6$ V. The lower figures show the simulation images of Ho^{14+} ions corresponding to the above plot. The vertical dotted line indicates the center of the trap along the z-axis and single Be$^+$ ion is located at the cross point. $\langle r \rangle$ value indicates the average radial position of the Ho^{14+} ions. The other simulation parameters are the same as in Fig. 3 except for V_z and the intensity of the cooling laser (10 mW/cm^2)

as in Fig. 3a except for the applied asymmetric field of $E_x = 5$ V/m. We observe that a considerable increase of the average kinetic energy of both Be$^+$ and Ho^{14+}. Since the positions of the ions shift from the trap axis by the static electric field, the average kinetic energy of Ho^{14+} drastically increases by micromotions. At the quasi-equilibrium state, the average radial position and the kinetic energy of Ho^{14+} are 4.3(2) μm and 2.4 K, respectively.

Next we tested the axial voltage dependence of sympathetic cooling efficiency. As shown in Fig. 4, the sympathetic cooling efficiency decreases with increasing V_z. By applying a higher V_z the distance between adjacent ions becomes shorter and the radial positions of the ions easily deviate from the trap axis by ion-ion collisions. Then the rf heating occurs on the HCIs, as mentioned in the previous section. As a result, the final average-kinetic energy of Ho^{14+} drastically increases.

It may be worth mentioning about the influence of initial conditions in MD simulations. Figure 5 shows the reproduce figure of the observed arrangement of the ion Coulomb crystal consisting of 10 Ho^{14+} and a single Be$^+$. A colored circle indicates an approximated position of each ion. These arrangements were obtained by randomly changing the initial positions of the ions in the MD simulations. As expected the final kinetic energy of Ho^{14+} ions possibly depends on the ion arrangement. In this point there is room for further investigation.

Finally we tried to increase the number of cold Ho^{14+} ions by sympathetic cooling with a few Be$^+$ ions. One of the simulation results is shown in Fig. 6. In this example, we performed the MD simulation of 16 Ho^{14+} and 3 laser-cooled Be$^+$ ions. The final average kinetic energy of Ho^{14+} is sufficiently lower than 1 mK at the quasi-equilibrium state. However, we anticipate that it is difficult to increase further the number of cold Ho^{14+} ions by increasing the number of Be$^+$ ions, because the increase of the space charge caused by Ho^{14+} and Be$^+$ ions easily induces the rf heating of the ions. The solution to this problem is the use of a segmented linear Paul trap. That is, it is possible to increase the number of cold Ho^{14+} ions up to over 100 by connecting 10 segmented linear Paul trap in series. This idea can be realized by applying the similar method described in Ref. [19] to a linear Paul trap.

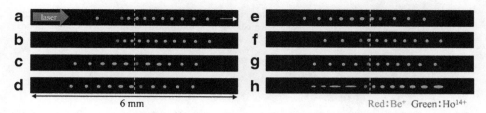

Fig. 5 Reproduce figures of the observed arrangements of the linear ion Coulomb crystal containing 10 Ho^{14+} (*green*) and a single Be^+ (*red*). The colored circle indicates the approximated position of each ion. The vertical dotted line shows the center of the trap

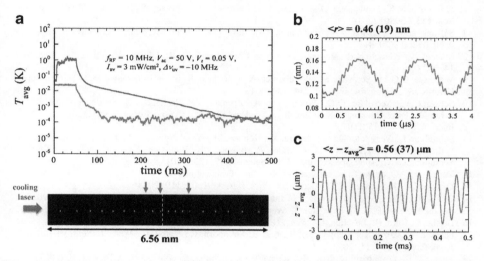

Fig. 6 Sympathetic cooling of 16 Ho^{14+} ions with 3 laser-cooled Be^+ ions. (a) Plots of the average kinetic energies of 16 Ho^{14+} and Be^+ ions. The final average kinetic energy of the Ho^{14+} ions is $1.3(8) \times 10^2$ μK. The lower image shows the ion arrangement in the quasi-equilibrium state. The red arrows in the image indicate the positions of the Be^+ ions. The simulation parameters are shown in the figure and the other parameters are the same as in Fig. 1. Figs (b) and (c) show example plots of the time variation of r and $z - z_{avg}$ of a certain Ho^{14+} ion. We also show the average values of r and $z - z_{avg}$ at the quasi-equilibrium state

4 Summary

In summary we have performed MD simulations to find out the conditions for efficient sympathetic cooling of HCIs in a linear Paul trap. Our extensive simulations show that the space charges of HCIs play important roles for the determinations of the axial distribution and the sympathetic cooling limit of the ions. Moreover, we found that a single laser cooled Be^+ can generate sufficiently cold at least 10 Ho^{14+} ions in a linear Paul trap with the ideal trapping fields, and the number of cold HCIs can increase using the segmented linear Paul trap in series. Since cold HCIs produced by sympathetic cooling are promising candidates for probing possible time variation of the fine structure constant [5–7, 13], further simulation studies will be performed in future.

Acknowledgements We thank to H. Katori for helpful comments and suggestions. This work is financially supported in part by a Grant-in-Aid for Young Scientists (A) from the Ministry of Education, Culture, Sports, Science and Technology (MEXT).

References

1. Okada, K., Suganuma, T., Furukawa, T., Takayanagi, T., Wada, M., Schuessler, H.A.: Phys. Rev. A **87**, 043427 (2013)
2. Hall, F.H.J., Aymar, M., Bouloufa-Maafa, N., Dulieu, O., Willitsch, S.: Phys. Rev. Lett. **107**, 243202 (2011)
3. Schiller, S., Korobov, V.: Phys. Rev. A 71, 032505 (2005)
4. Kajita, M., Gopakumar, G., Abe, M., Hada, M., Keller, M.: Phys. Rev. A 89, 032509 (2014)
5. Berengut, J.C., Dzuba, V.A., Flambaum, V.V.: Phys. Rev. Lett. 105, 120801 (2010)
6. Safronova, M.S., Dzuba, V.A., Flambaum, V.V., Safronova, U.I., Porsev, S.G., Kozlov, M.G.: Phys. Rev. Lett. **113**, 030801 (2014)
7. Safronova, M.S., Dzuba, V.A., Flambaum, V.V., Safronova, U.I., Porsev, S.G., Kozlov, M.G.: Phys. Rev. A **90**, 052509 (2014)
8. Mäckel, V., Klawitter, R., Brenner, G., Crspo López-Urrutia, J.R., Ullrich, J.: Phys. Rev. Lett. **107**, 143002 (2011)
9. Versolato, O.O.M., Schwarz, M., Windberger, A., Ullrich, J., Schmidt, P.O., Drewsen, M., Crspo López-Urrutia, J.R.: Hyperfine Interact. **214**, 189 (2013)
10. Schwarz, M., Versolato, O.O., Windberger, A., Brunner, F.R., Ballance, T., Eberle, S.N., Ullrich, J., Schimidt, P.O., Hansen, A.K., Gingell, A.D., Drewsen, M., Crespo López-Urrutia, J.R.: Rev. Sci. Instrum. **83**, 083115 (2012)
11. Schmöger, L., Versolato, O.O., Schwarz, M., Kohnen, M., Windberger, A., Piest, B., Feuchtenbeiner, S., Pedregosa-Gutierrez, J., Leopold, T., Micke, P., Hansen, A.K., Baumann, T.M., Drewsen, M., Ullrich, J., Schmidt, P.O., Crespo López-Urrutia, J.R.: Science **347**, 1233 (2015)
12. Versolato, O.O. et al.: To be appeared in Hyp. Int. of this volume
13. Dzuba, V.A., Flambaum, V.V., Katori, H.: Phys. Rev. A 91, 022119 (2015)
14. Okada, K., Wada, M., Takayanagi, T., Ohtani, S., Schuessler, H.A.: Phys. Rev. A **81**, 013420 (2010)
15. Okada, K., Yasuda, K., Takayanagi, T., Wada, M., Schuessler, H.A., Ohtani, S.: Phys. Rev. A **75**, 2007
16. Blythe, P., Roth, B., Fröhlich, U., Wenz, H., Schiller, S.: Phys. Rev. Lett. **95**, 183002 (2005)
17. Feldker, T., Pelzer, L., Stappel, M., Bachor, P., Steinborn, R., Kolbe, D., Walz, J., Schmidt-Kaler, F.: Appl. Phys. B **114**, 11 (2014)
18. Zhang, C.B., Offenberg, D., Roth, B., Wilson, M.A., Schiller, S.: Phys. Rev. A **76**, 012719 (2007)
19. Fujitaka, S., Wada M., et al.: Nucl. Instrum. M.th. B **126**, 386 (1997)

Hyperfine Interact (2015) 236:95–100
DOI 10.1007/s10751-015-1206-0

Laser spectroscopy of atoms in superfluid helium for the measurement of nuclear spins and electromagnetic moments of radioactive atoms

T. Fujita[1] · T. Furukawa[2] · K. Imamura[3,4] · X. F. Yang[3,5] · A. Hatakeyama[6] ·
T. Kobayashi[7] · H. Ueno[3] · K. Asahi[8] · T. Shimoda[1] · Y. Matsuo[9] ·
OROCHI Collaboration

Published online: 26 November 2015
© Springer International Publishing Switzerland 2015

Abstract A new laser spectroscopic method named "OROCHI (Optical RI-atom Observation in Condensed Helium as Ion catcher)" has been developed for deriving the nuclear spins and electromagnetic moments of low-yield exotic nuclei. In this method, we observe atomic Zeeman and hyperfine structures using laser-radio-frequency/microwave double-resonance spectroscopy. In our previous works, double-resonance spectroscopy was performed successfully with laser-sputtered stable atoms including non-alkali Au atoms as well as alkali

---

Proceedings of the 6th International Conference on Trapped Charged Particles and Fundamental
Physics (TCP 2014), Takamatsu, Japan, 1–5 December 2014

✉ T. Fujita
 tomomi.fujita@riken.jp

1   Department of Physics, Osaka University, 1–1 Machikaneyama, Toyonaka, Osaka 560–0043, Japan

2   Department of Physics, Tokyo Metropolitan University, 1–1 Minami-Osawa, Hachioji, Tokyo
    192–0397, Japan

3   RIKEN Nishina Center, 2–1 Hirosawa, Wako, Saitama 351–0198, Japan

4   Department of Physics, Meiji University, 1-1-1 Higashimita, Tama, Kawasaki, Kanagawa
    214-8571, Japan

5   School of Physics, Peking University, Chengfu Road, Haidian District, Beijing, 100871, China

6   Department of Applied Physics, Tokyo University of Agriculture and Technology, 2–24–16
    Naka-cho, Koganei, Tokyo 184–8588, Japan

7   RIKEN Center for Advanced Photonics, 2–1 Hirosawa, Wako, Saitama 351–0198, Japan

8   Department of Physics, Tokyo Institute of Technology, 2–12–1 Oh-Okayama, Meguro, Tokyo
    152–8551, Japan

9   Department of Advanced Sciences, Hosei University, 3–7–2 Kajino-cho, Koganei-shi, Tokyo,
    184–8548, Japan

Rb and Cs atoms. Following these works, measurements with $^{84-87}$Rb energetic ion beams were carried out in the RIKEN projectile fragment separator (RIPS). In this paper, we report the present status of OROCHI and discuss its feasibility, especially for low-yield nuclei such as unstable Au isotopes.

Keywords Laser spectroscopy in superfluid helium · Double-resonance spectroscopy · Optical pumping · Nuclear spin · Electromagnetic moments · Hyperfine interaction

1 Introduction

Nuclear spins and electromagnetic moments are key observables for investigating nuclear structures as they are sensitive to the configurations of valence nucleons. Laser spectroscopy is an effective tool for understanding nuclear structure through the determination of those observables [1]. However, the application of these methods to far-unstable rare isotopes, for which anomalous nuclear properties have so far been reported, is still limited mainly due to various technical difficulties arising from the low production yield of rare isotopes. Hence, we have developed a new laser spectroscopic method named OROCHI (Optical Radioisotope atom Observation in Condensed Helium as Ion catcher) for low-yield RI atoms. In this method, atomic Zeeman and hyperfine structure splittings are measured using a laser-radio frequency (RF)/microwave (MW) double-resonance method. The significant feature of the OROCHI method is the utilization of superfluid helium (He II) as both an effective trapping material and a host matrix for laser spectroscopy taking advantage of characteristic properties of atoms in He II [2]. We have started performing measurements of not only stable alkali Rb and Cs atoms but also non-alkali Au atoms. Furthermore, we have successfully applied this technique to $^{84-87}$Rb energetic ion beams. Here, we report on the present status of OROCHI for measurements with laser-sputtered atoms and with energetic ion beams. Then, we discuss the feasibility of OROCHI for the measurement of nuclei far from stability, which will be performed under changing conditions, such as high ion injection energy at low yield.

2 OROCHI–new laser spectroscopic method for studying RIs

In the OROCHI method, energetic ion beams are injected into He II. These ion beams are decelerated and finally stopped as neutralized atoms via the capture of free electrons. He II can be utilized to stop almost all the atoms in the observation region (2×5 mm^2) owing to its high stopping power. The injected atoms reside in the observation region for a sufficiently long period owing to the slow diffusion of atoms in He II (typically a few mm/s). Moreover, no macroscopic bubbles appear because He II evaporates only from the surface [3].

The trapped atoms are irradiated with a pumping laser and emit laser-induced fluorescence (LIF) photons. The absorption lines of atoms in He II are blue shifted and considerably broadened compared with those in vacuum. This is caused by the interaction with the surrounding helium atoms. These characteristic properties enable us to perform the measurement with high signal-to-noise ratio by reducing the background photon count. By taking advantage of the difference between the wavelengths of absorption and emission lines, a wavelength separation device such as an interference filter or a monochromator can be utilized to efficiently reduce the detection of stray laser light, which is the main source of

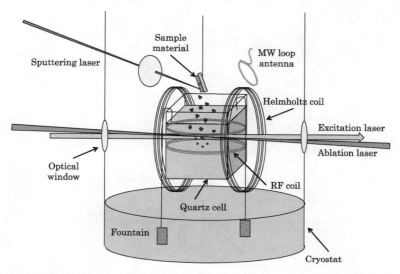

Fig. 1 Schematic diagram of the measurements with laser-sputtered atoms. In a cryostat, An open-topped cubic quartz cell ($7 \times 7 \times 7$ cm^3) filled with He II at a temperature of approximately 1.6 K was placed in a cryostat. The sample material was placed 1 cm above the He II surface. Around the quartz cell, Helmholtz coils, RF coils and an MW loop antenna were installed to perform double-resonance spectroscopy. The applied external magnetic field was typically 1 G. The emitted LIF photons were observed using a monochromator and a PMT

the background count. We then obtain the atomic Zeeman and hyperfine splittings in He II by the laser-RF/MW double-resonance method, efficiently [4].

3 Double-resonance spectroscopy using laser-sputtered atoms in He II

We have measured the Zeeman and hyperfine splittings of Rb, Cs and Au. Figure 1 shows the experimental setup. An open-topped cubic quartz cell ($7 \times 7 \times 7$ cm^3) in a cryostat was fully filled with He II by use of the superfluid fountain effect. The temperature of He II was typically maintained at 1.6 K. The sample material was placed 1 cm above the He II surface and ablated by a second- or third-harmonic pulse of a Nd:YAG (yttrium aluminum garnet) laser (wavelength: 355 or 532 nm, pulse duration: 8 ns). Most of the sputtered particles, which entered He II, were formed as clusters. To dissociate the clusters, we irradiated them with a femtosecond Ti:sapphire laser (wavelength: 800 nm, repetition rate: 500 Hz) [5]. Helmholtz coils, RF coils and an MW loop antenna were installed around the quartz cell to perform optical pumping and laser-RF/MW double-resonance spectroscopy. The applied external magnetic field was typically 1 G. LIF photons emitted from laser-excited atoms were focused with three lenses and the wavelength was separated by a monochromator and detected by a photomultiplier tube (PMT).

The LIF intensity was decreased when atoms were polarized by the irradiation of a circularly polarized pumping laser light. Then, the atomic spin polarizations were determined from the ratio of LIF intensities observed by irradiating with linearly and circularly polarized lasers. In our measurements, large atomic spin polarizations were confirmed using alkali Cs atoms ($\simeq$ 90 %) and Rb atoms ($\simeq$ 50 %) in He II, respectively. Recently, the spin

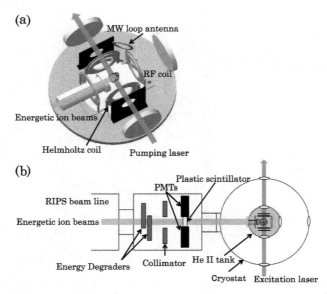

Fig. 2 Setup of expiredent at the RIPS ion beam line. **a** Inside of the cryostat filled with He II at a temperature of approximately 1.5 K. Helmholtz coils, RF coils and an MW loop antenna were installed to perform RF/MW double-resonance spectroscopy. **b** System used to optimize beam-stop position. Two aluminum degraders were installed upstream of the cryostat to adjust the beam energy. A plastic scintillator upstream of the cryostat detected the intensity of injected ions

polarization of Au atoms has also been successfully achieved using a pulsed laser light, here the degree of atomic spin polarization was larger than 80 %.

Using the large atomic spin polarization, we also succeeded in measuring Zeeman and hyperfine resonances for stable Rb, Cs and Au atoms. Further analysis of the experimental data is in progress.

4 Experiment using energetic ion beams

We have also performed measurements using $^{84-87}$Rb energetic ion beams with energies of 60–66 MeV/u at the RIKEN projectile fragment separator (RIPS) at the RIKEN Radioactive Isotope Beam Factory (RIBF) [6, 7]. Figure 2a shows the inside of the cryostat for the ion beam experiment (fulfilled with He II at a temperature of approximately 1.5 K). The energetic ion beams from the RIPS beam line were injected into the cryostat (horizontal arrow in Fig. 2b) and the stopped RI atoms were subjected to a pumping laser (vertical arrow in Fig. 2b, cw Ti: sapphire laser, wavelength: 780 nm, laser power: ~120 mW, beam diameter: typically 2 mm). Beneath the cryostat, a photodetection system for observing LIF photons, which included three lenses, an interference filter and a Peltier-cooled PMT were installed. We performed a laser RF/MW double-resonance spectroscopy measurement with the RI atoms stopped in He II. One of the crucial points in this measurement was the optimization of the beam-stop position. Only the photon signals from the observation region (the center of the region shown in Fig. 2a) were focused and detected. To the optimize the beam stop position, two Al degraders and a plastic scintillator were installed upstream of the cryostat, as shown in Fig. 2b. We adjusted the thickness of the Al degraders from

Table 1 Nuclear spins derived in previous works

| | | Nuclear spin | | |
Isotopes		Our result	Literature value	Ref.
^{84}Rb		1.9(1)	2	[8]
^{84m}Rb		6.2(2)	6	[8, 9]
^{85}Rb		2.5(1)	5/2	[8]
^{86}Rb		1.9(2)	2	[8]
^{87}Rb		1.53(6)	3/2	[8]

Details of experimental results and the discussion is in reference [2]

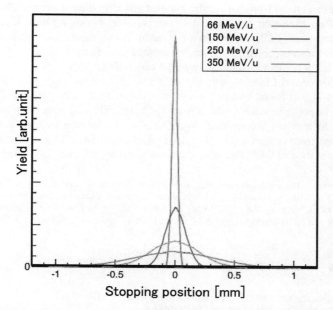

Fig. 3 LISE++ calculation result for the spread of the stopping range for beam energies of 66, 150, 250 and 350 MeV/u [11]. The previous measurements at the RIPS beam line were performed with a beam energy of 66 MeV/u [6]. Note that we did not take into account the energy distribution upstream of the cryostat

0 to 800 μm with a 12.5 μm step to vary the beam energy. The plastic scintillator was used to count the number of injected ions. The beam-stop position was estimated by counting the LIF photons from atoms while changing the degrader thickness.

Zeeman resonance frequencies were determined for $^{84-87}$Rb. The intensity in the measurements were on the order of 10^4 pps, and the FWHM of the stop range was approximately 1 mm. The deduced nuclear spins were consistent with the literature values as shown in Table 1 [2, 8, 9].

For the future application of OROCHI, to nuclei far from stability, we plan to perform measurements at BigRIPS at RIBF, where the energy of ions becomes as high as 345 MeV/u [10]. In this case, the stopping range should be considered and optimized carefully. Figure 3 shows the result of a LISE++ calculation [11] estimating the stopping distribution for ^{87}Rb primary beams of 66, 150, 250 and 350 MeV/u. The FWHMs are 0.05 mm for the 66 MeV/u beam, 0.12 mm for a 150 MeV/u beam, 0.48 mm for a 250 MeV/u beam and 0.82 mm for a

350 MeV/u beam. All the FWHMs are within 1 mm. We conclude that the area can be fully covered by the pumping laser.

5 Conclusion

A new laser spectroscopic method named OROCHI (Optical RI-atom Observation in Condensed Helium as Ion catcher) has been developed for the investigation of nuclear spins and electromagnetic moments of low-yield exotic nuclei. In this method, we utilize He II as both an effective trapping material and a host matrix for laser spectroscopy by taking advantage of the characteristic properties of atoms in He II. Nuclear spins and electromagnetic moments can be derived from the atomic Zeeman and hyperfine structures observed by the laser-RF/MW double-resonance method. In our previous works, we have successfully performed measurements on laser-sputtered non-alkali Au atoms as well as alkali Rb and Cs atoms. We have also succeeded in observing atoms from injected energetic $^{84-87}$Rb beams produced by RIPS with a beam intensity on the order of 10^4 pps.

We have performed LISE++ calculations pertinent to future applications with higher-energy beams. It was found that in the case of measurements with 350 MeV/u beams, the straggling of stopping position in He II is within 1 mm. The pumping laser can fully cover the area in which atoms are stopped. Note that we did not take into account the energy distribution upstream of the cryostat. It will be necessary to perform further calculations for the future application of OROCHI to exotic lower-yield nuclei such as unstable Au isotopes.

Acknowledgments The measurement using energetic ion beams was performed under Program No. NP0802-RRC53 at RIBF, operated by RIKEN Nishina Center and Center for Nuclear Study (CNS), The University of Tokyo. We thank the RIKEN Ring Cyclotron staff for their cooperation during the measurement. This work was partly supported by KAKENHI through a Grant-in-Aid for Scientific Research.

References

1. Kluge, H., Nörtershäuser, W.: Spectrochim. Acta B **58**, 1031 (2003)
2. Yang, X.F., et al.: Phys. Rev. A **90**, 052516 (2014)
3. Fujisaki, A., et al.: Phys. Rev. Lett. **71**(7), 1039 (1993)
4. Lang, S., et al.: Europhys. Lett. **30**(4), 233 (1995)
5. Furukawa, T., et al.: Phys. Rev. Lett. **96**, 095301 (2006)
6. Kubo, T., et al.: Nucl. Instrum. Methods B **70**, 309 (1992)
7. Yang, X.F., et al.: Nucl. Instrum. Methods B **317**, 599 (2013)
8. Thibault, C., Touchard, F., Buttgenbach, S., et al.: Phys. Rev. C **23**, 2720 (1981)
9. Mazur, V.M., Bigan, Z.M., Symochko, D.M.: J. Phys. G: Nucl. Part. Phys. **37**, 035101 (2010)
10. Kubo, T.: Nucl. Instrum. Methods B **204**, 97 (2003)
11. Tarsal, O.B., Bazin, D.: Nucl. Instrum. Methods B **266**, 4657 (2008)

Hyperfine Interact (2015) 235:13–20
DOI 10.1007/s10751-015-1205-1

The ASACUSA CUSP: an antihydrogen experiment

N. Kuroda[1] · S. Ulmer[2] · D. J. Murtagh[3] · S. Van Gorp[3] · Y. Nagata[3,10] ·
M. Diermaier[4] · S. Federmann[5] · M. Leali[6] · C. Malbrunot[5] · V. Mascagna[6] ·
O. Massiczek[4] · K. Michishio[7] · T. Mizutani[1] · A. Mohri[8] · H. Nagahama[1] ·
M. Ohtsuka[1] · B. Radics[3] · S. Sakurai[9] · C. Sauerzopf[4] · K. Suzuki[4] ·
M. Tajima[1] · H. A. Torii[1] · L. Venturelli[6] · B. Wünschek[4] · J. Zmeskal[4] ·
N. Zurlo[6] · H. Higaki[9] · Y. Kanai[3] · E. Lodi Rizzini[6] · Y. Nagashima[7] ·
Y. Matsuda[1] · E. Widmann[4] · Y. Yamazaki[3]

Published online: 10 November 2015
© Springer International Publishing Switzerland 2015

Abstract In order to test CPT symmetry between antihydrogen and its counterpart hydrogen, the ASACUSA collaboration plans to perform high precision microwave spectroscopy of ground-state hyperfine splitting of antihydrogen atom in-flight. We have developed an apparatus ("cusp trap") which consists of a superconducting anti-Helmholtz coil and

Proceedings of the 6th International Conference on Trapped Charged Particles and Fundamental
Physics (TCP 2014), Takamatsu, Japan, 1–5 December 2014.

✉ N. Kuroda
kuroda@phys.c.u-tokyo.ac.jp

[1] Graduate School of Arts and Sciences, University of Tokyo, 3-8-1 Komaba Meguro-ku,
     153-8902, Tokyo, Japan

[2] Ulmer Initiative Research Unit, RIKEN, 2-1 Hirosawa Wako-shi, 351-0198, Saitama, Japan

[3] Atomic Physics Laboratory, RIKEN, 2-1 Hirosawa Wako-shi, 351-0198, Saitama, Japan

[4] Stefan Meyer Institut für Subatomare Physik, Boltzmangasse 3, 1090 Wien, Austria

[5] CERN, CH-1211, Genève, Switzerland

[6] Dipartimento di Chimica e Fisica per l'Ingegneria e per i Materiali, Università di Brescia
     & Instituto Nazionale di Fisica Nucleare, Gruppo Collegato di Brescia, 25133 Brescia, Italy

[7] Department of Physics, Tokyo University of Science, Kagurazaka Shinjuku-ku,
     162-8601, Tokyo, Japan

[8] Graduate School of Human and Environmental Sciences, Kyoto University, Yoshida
     Nihonmatsu-cho Sakyo-ku, 606-8501, Kyoto, Japan

[9] Graduate School of Advanced Science of Matter, Hiroshima University, Kagamiyama,
     Higashi-Hiroshima, 739-8530, Hiroshima, Japan

[10] Present address: Department of Applied Physics, Tokyo University of Agriculture and Technology,
     Naka-cho Koganei-shi, 184–8588, Tokyo, Japan

multiple ring electrodes. For the preparation of slow antiprotons and positrons, Penning-Malmberg type traps were utilized. The spectrometer line was positioned downstream of the cusp trap. At the end of the beamline, an antihydrogen beam detector was located, which comprises an inorganic Bismuth Germanium Oxide (BGO) single-crystal scintillator housed in a vacuum duct and surrounding plastic scintillators. A significant fraction of antihydrogen atoms flowing out the cusp trap were detected.

Keywords Antihydrogen · CPT invariance · Atomic beam · Rydberg atom

1 Introduction

Using slow antiprotons from the CERN AD (Antiproton Decelerator), the ASACUSA (Atomic Spectroscopy And Collisions Using Slow Antiprotons) collaboration studies fundamental atomic processes such as ionization by antiproton collisions, antiprotonic atom formation, and antiproton annihilation, as well as the structure of antiprotonic atoms and antihydrogen. Antihydrogen, the antipartner of hydrogen, is an excellent target to perform stringent tests of CPT symmetry. The ASACUSA CUSP experiment plans spectroscopic studies of the ground-state hyperfine splitting of antihydrogen atoms in-flight [1].

The idea of the hyperfine spectroscopy experiment is based on classical Rabi-like atomic beam magnetic-resonance detection method. We have developed the experimental apparatus (Fig. 1) in order to accomplish this aim. The constituent particles of antihydrogen, the antiproton and positron, have relatively high kinetic energies when they are produced. The accelerator-generated antiprotons were confined and cooled in a Penning-Malmberg type trap (the MUSASHI trap). Positrons from a radio-active β^+ source were moderated, cooled, and accumulated in another trap. The most important part of the setup is the cusp trap where antihydrogen atoms were synthesized. Pion tracking detectors were placed on both sides of the cusp trap, and the flow of antihydrogen atoms detected at the end of the spectrometer line.

2 MUSASHI, an ultraslow antiproton beam source

The MUSASHI trap serves for capture antiprotons from the CERN AD and cooled them down to sub-eV energy. The stored antiprotons can then be extracted as a 150 eV ultraslow beam [2].

The 5.3 MeV antiproton beam from the AD is generally slowed down via an energy degrader [3] with a area density of 70 mg/cm^2. In the case of the ASACUSA CUSP experiment this simple degrader foil is replaced by a radio frequency quadrupole decelerator (RFQD) [4], and relatively thin foils (180 μg/cm^2 in total).

The RFQD decelerated 5.3 MeV antiproton beams with an efficiency of typically 30 %. The potential of the entire RFQD was able to be floated by 60 kV, the value of this floating is tuned so as to maximize the capture efficiency of the MUSASHI trap. This occurred at a beam energy of 111.5 keV.

By passing through the thin degrader foils, the antiprotons suffered a finite energy loss reducing the longitudinal energy of the antiproton beam to less than 12 keV.

The MUSASHI trap is comprised from multiple ring electrodes (MRE). The MRE generates an axisymmetric potential which compensated the self-charge induced potential of

Fig. 1 Setup of the ASACUSA CUSP experiment

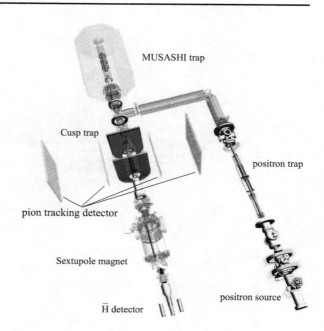

MUSASHI trap

Cusp trap

positron trap

pion tracking detector

Sextupole magnet

positron source

$\bar{H}$ detector

the confined plasma [6]. Due to this, a stable confinement of non-neutral plasma was achieved. The injected antiproton beam is captured inside the MRE and cooled via collisions with a pre-loaded electron plasma. Since the MRE is housed in a 2.5 T superconducting solenoid, the electrons dissipate their energy via cyclotron radiation. Thus, the antiprotons are cooled.

After electron cooling, the antiproton cloud which has a similar radius to the electron plasma ($r = 3.5$ mm) [8] requires radial compression for a high extraction efficiency. Prior to the compression, the electrons are removed from the trap region by opening the potential for a short time which is long enough to lose electrons but short enough to keep antiprotons (ca. 500 ns). The cooled antiproton cloud was radially compressed by applying a torque using a rotating electric field via a four-fold segmented electrode [7, 8]. The compressed antiproton cloud with a diameter of 0.5 mm [8] and an axial length of 1.6 mm in a potential well was kicked out. The extracted 150 eV beams are guided by magnetic coils. The pulse length became around 2 μs at the cusp trap.

3 Preparation of slow positrons

Positrons are obtained from a radioactive ^{22}Na source instead of accelerators. The positron system in 2012 was upgraded by replacing tungsten moderators for a solid Ne moderator [10, 11]. The particles are moderated by a layer of solid Ne grown on a small conical structure in front of the source. The emitted positrons are guided by magnetic coils and introduced into a trap (the positron accumulator shown in Fig. 1). By interaction with N_2 gas, the positrons were decelerated and accumulated in a electro-static potential well in a 0.3 T magnetic field environment. Subsequently, the accumulated positrons were transferred to the cusp trap. By repeating this sequence 30 times, 3×10^7 positrons were loaded in the cusp trap.

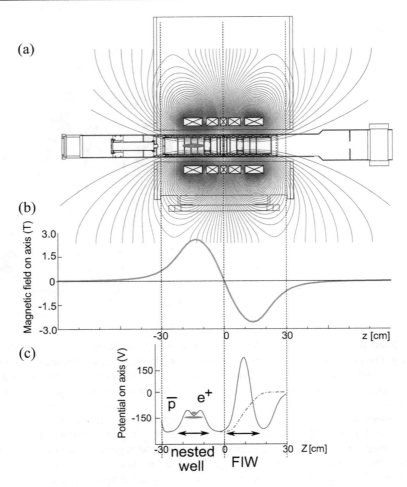

Fig. 2 a A cross sectional view of the cusp trap and a magnetic field lines. **b** The cusp magnetic field along the axis. **c** *Solid line*: the electric potential along the axis to mix p̄ with e^+. *Dashed line*: to dump p̄s accumulated in the field-ionization well (FIW)

4 Antihydrogen synthesis in the cusp trap

To synthesize antihydrogen atoms, the 150 eV antiproton beam from the MUSASHI trap was directly injected into the positron plasma confined in the cusp trap.

The cusp trap consisted of superconducting anti-Helmholtz coils and multiple ring electrodes. Figure 2a shows a schematic cross sectional view of the cusp trap with its magnetic field lines. The magnetic field along the axis is shown in Fig. 2b. The MRE was located in an ultrahigh vacuum bore which is kept at 6 K by two cryocoolers [9]. On both sides of the bore, apertures which act as thermal shields are installed in order to reduce thermal flow into the 6 K cold region.

A nested well configuration was prepared using the MRE as shown in Fig. 2c. Positrons were loaded into the center of the nested well, which was placed in the relatively high

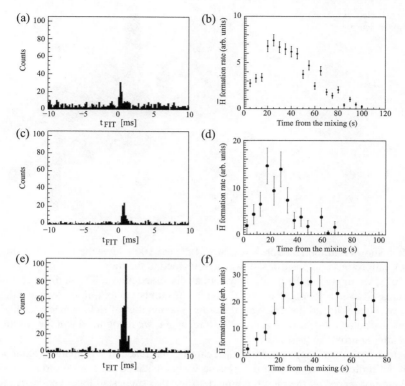

Fig. 3 **a** A spectrum of antiproton annihilation when the field-ionization trap was dumped. **b** The number of field-ionized H̄ atoms monitored by opening the FIW every 5 s, which were normalized to one cycle (19 cycles were taken). **c** and **d** were obtained in 2012, when we increased the number of positrons. **e** and **f** taken from 1 typical cycle were obtained when an rf-field was applied to enhance the yield of antihydrogen atoms

(2.7 T) magnetic field region. The field reduced by 5 % at the edge of the positron plasma. The density of positron plasma was estimated at the order of $10^8 \mathrm{cm}^{-3}$, and the temperature to be the order of 10^2 K [12]. The kinetic energy of the antiproton beam, 150 eV, was slightly above the potential energy of the positron plasma. Antiprotons from the MUSASHI trap collided with positrons. Though positrons were heated up by the injected beam, they cooled via cyclotron radiation. Antiprotons underwent sympathetic cooling and eventually captured a positron, thus an antihydrogen atom was formed.

A field-ionization well (FIW) shown in Fig. 2c was prepared to detect antihydrogen atoms formed in the cusp trap region. Synthesized antihydrogen moved freely because of its electrical neutrality. Some fraction of them reached the FIW and could be ionized if they were in a Rydberg state [13]. Antiprotons from the field-ionization of antihydrogen were dumped to annihilate on the surrounding wall. Thus pions were produced and then counted by the tracking detector.

In 2010, we succeeded in synthesizing antihydrogen atom by injecting 3×10^5 antiprotons into 3×10^6 positrons, which was confirmed with the field ionization technique [12]. The measured time spectrum is shown in Fig. 3a, while the time evolution of the number of field-ionized antihydrogen atom was Fig. 3b. Typically, the number of field-ionized antihydrogen atoms after correction of the solid angle of the pion detector was 70. The time

Fig. 4 A 3D view of the antihydrogen detector

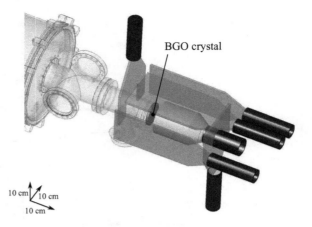

BGO crystal

10 cm 10 cm
10 cm

evolution shows that the reaction was almost finished 100 s after the antiproton injection. Most of antiprotons and positrons, however, remained in the trap.

In 2012, we upgraded the positron system as discussed above. The number of positrons was increased by a factor of 10, ie. 3×10^7, while the number of antiproton was kept the same as before. Figure 3c and d show typical results. The number of field-ionized antihydrogen atom increased by a factor of 2.7, ie. 190. Here, again, we recognized that the reaction had ceased after approximately 40 s.

In order to explain these phenomena regarding the reaction termination, axial separation of antiprotons from the positron plasma was considered. The antiproton annihilation distribution observed by the pion tracking detector also suggested this hypothesis [12]. To counteract the axial separation and to prolong the reaction, an rf field was applied. If the frequency of the rf was tuned to excite the axial oscillation of the separated antiprotons, the yield of field-ionized antihydrogen atom was improved by a factor of 3.5 and the reaction was continued for longer as shown in Fig. 3e and f. Hence, approximately 660 antihydrogen atoms per one mixing cycle were detected [14], which is almost 10 times more than that reported in 2010 [12].

5 Antihydrogen beam detection

In order to detect antihydrogen atoms which flow out from the cusp trap, an antihydrogen detector was installed at the end of the spectrometer line, 2.7 m downstream from the center of the cusp trap. At 1.5 m downstream from the center, where a microwave cavity will be installed, the residual magnetic fields were already low enough to perform a spectroscopic measurement.

Figure 4 shows the 3D schematic drawing of the antihydrogen detector which is comprised of an inorganic BGO single crystal scintillator and five plastic scintillators. The BGO crystal has a diameter of 100 mm and a thickness of 5 mm. It was installed inside a vacuum chamber with its center on the beam axis, hence antihydrogen atoms may hit the crystal directly. To detect charged pions by annihilation, the five plastic scintillator plates with thickness of 10 mm were placed outside of the chamber. The coverage of the plastic scintillators was 49 % of 4π. Each scintillator is read out by a photomultiplier tube (PMT). The signal of the BGO scintillator is recorded by a fast waveform digitizer. The

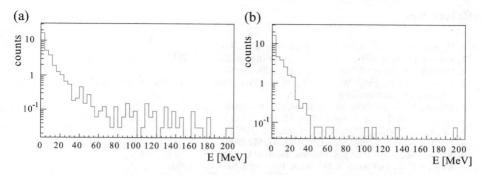

Fig. 5 Energy deposition on the BGO. **a** When antiprotons were mixed with positrons. **b** Antiprotons were trapped in the cusp trap without positrons

timing information of all scintillators is read out by time-to-digital converters. A set of field-ionization (FI) electrodes was located in front of the BGO scintillator in order to investigate the principal quantum number of $\bar{H}$ atoms reaching the detector.

Figure 5a shows the measured energy deposition on the BGO crystal when 3×10^5 antiprotons were mixed with 3×10^7 positrons in the presence of rf-assistance. In contrast, Fig. 5b shows the results when antiprotons were trapped without positrons. An excess can be seen at deposition energies above 40 MeV, consistent with a Monte Carlo (GEANT4) simulation of the detector performance. The number of antihydrogen atoms was evaluated as 25 and 16 per hour for different field strength on the FI electrodes of 94 and 452 V/cm, respectively [14]. This sets an upper limit on the principal quantum number of the quantum state of the detected antihydrogen atoms of 43 for 94 V/cm and 29 for 452 V/cm. The statistical significance in terms of Gaussian standard deviations by taking the ratio of Poisson means indicates 4.8σ for $n = 43$ and 3.0σ for $n = 29$ [14].

6 Summary and outlook

For a stringent test of CPT symmetry, the ASACUSA collaboration has developed the cusp trap, the MUSASHI trap, the positron accumulator, the spectrometer line, and the antihydrogen detector. We have demonstrated the successful synthesis of cold antihydrogen atoms employing the cusp trap, and succeeded in detecting antihydrogen atoms where residual magnetic fields were small.

We are in process of performing the microwave spectroscopy of antihydrogen by upgrading the cusp magnet for improved extraction and focusing efficiency, improving the antiproton beam quality for efficient antihydrogen synthesis, developing a new tracking detector for monitoring purpose of antihydrogen synthesis, and upgrading the antihydrogen detector for a better S/N ratio. The details of these upgrades and the physics results will be reported soon.

Acknowledgements We would wish to express our gratitude towards the PS, AD, and RF groups at CERN. This work was supported by the Grant-in-Aid for Specially Promoted Research (19002004 and 24000008) of Japanese Ministry of Education, Culture, Sports, Science and Technology (Monbukagaku-shō), Special Research Projects for Basic Science of RIKEN, RIKEN FPR program, RIKEN IRU program, European Research Council under European Union's Seventh Framework Programme (FP7/2007–2013)/ERC Grant agreement (291242), the Austrian Ministry for Science and Research, Università di Brescia, and Instituto Nazionale di Fisica Nucleare.

References

1. Widmann, E., et al.: Lect. Notes Phys. **570**, 528 (2001)
2. Kuroda, N., et al.: Phys. Rev. ST Accel. Beams **15**, 0247021 (2012)
3. Gabrielse, G., et al.: Phys. Rev. Lett. **57**, 2504 (1986)
4. Lombardi, A.M., Pirkl, W., Bylinsky, Y.: In: Proceedings of the 2001 Particle Accelerator Conference, Chicago, (IEEE, Piscataway, p. 585 (2001)
5. Kuroda, N., et al.: Phys. Rev. Lett. **94**, 023401 (2005)
6. Mohri, A., et al.: Jpn. J. Appl. Phys. **37**, 664 (1998)
7. Huang, X.-P., et al.: Phys. Rev. Lett. **78**, 875 (1997)
8. Kuroda, N., et al.: Phys. Rev. Lett. **100**, 203402 (2008)
9. Shibata, M., et al.: Rev. Sci. Instrum. **79**, 015112 (2008)
10. Mills, A.P. Jr., Gullikson, E.M.: Appl. Phys. Lett. **49**, 1121 (1986)
11. Murtagh D.J., et al. to be published
12. Enomoto, Y., et al.: Phys. Rev. Lett. **105**, 243401 (2010)
13. Gabrielse, G., et al.: Phys. Rev. Lett. **89**, 213401 (2002)
14. Kuroda, N., et al.: Nat. Commun. **5**, 3089 (2014)

Hyperfine Interact (2015) 235:21–28
DOI 10.1007/s10751-015-1202-4

Antiproton cloud compression in the ALPHA apparatus at CERN

A. Gutierrez[1] · M. D. Ashkezari[2] · M. Baquero-Ruiz[3] · W. Bertsche[4,5] ·
C. Burrows[6] · E. Butler[7,8] · A. Capra[9] · C. L. Cesar[10] · M. Charlton[6] ·
R. Dunlop[2] · S. Eriksson[6] · N. Evetts[1] · J. Fajans[3] · T. Friesen[11] ·
M. C. Fujiwara[12] · D. R. Gill[12] · J. S. Hangst[11] · W. N. Hardy[1] · M. E. Hayden[2] ·
C. A. Isaac[6] · S. Jonsell[13] · L. Kurchaninov[12] · A. Little[3] · N. Madsen[6] ·
J. T. K. McKenna[12] · S. Menary[9] · S. C. Napoli[13] · P. Nolan[14] · K. Olchanski[12] ·
A. Olin[12] · P. Pusa[14] · C. Ø. Rasmussen[11] · F. Robicheaux[15] ·
R. L. Sacramento[10] · E. Sarid[16] · D. M. Silveira[10] · C. So[3] · S. Stracka[12] ·
J. Tarlton[7] · T. D. Tharp[11] · R. I. Thompson[17] · P. Tooley[4] · M. Turner[3] ·
D. P. van der Werf[6] · J. S. Wurtele[3] · A. I. Zhmoginov[3]

Published online: 12 October 2015

Abstract We have observed a new mechanism for compression of a non-neutral plasma, where antiprotons embedded in an electron plasma are compressed by a rotating wall drive at a frequency close to the sum of the axial bounce and rotation frequencies. The radius of the antiproton cloud is reduced by up to a factor of 20 and the smallest radius measured is ~ 0.2 mm. When the rotating wall drive is applied to either a pure electron or pure antiproton plasma, no compression is observed in the frequency range of interest. The frequency range over which compression is evident is compared to the sum of the antiproton bounce

---

Proceedings of the 6th International Conference on Trapped Charged Particles and Fundamental Physics (TCP 2014), Takamatsu, Japan, 1–5 December 2014

✉ A. Gutierrez
andrea.gutierrez@triumf.ca

1 Department of Physics and Astronomy, University of British Columbia, Vancouver, British Columbia V6T 1Z1, Canada

2 Department of Physics, Simon Fraser University, Burnaby, British Columbia, V5A 1S6, Canada

3 Department of Physics, University of California at Berkeley, Berkeley, CA 94720-7300, USA

4 School of Physics and Astronomy, University of Manchester, M13 9PL Manchester, UK

5 The Cockcroft Institute, WA4 4AD Warrington, UK

6 Department of Physics, College of Science, Swansea University, Swansea SA2 8PP, UK

7 Centre for Cold Matter, Imperial College, London SW7 2BW, UK

frequency and the system's rotation frequency. It is suggested that bounce resonant transport is a likely explanation for the compression of antiproton clouds in this regime.

Keywords Antiprotons · Rotating wall · Compression · Electrons · Penning-Malmberg trap · Non-neutral plasma · Antihydrogen

1 Introduction

Antihydrogen is the simplest neutral antimatter atom. Precision comparisons between hydrogen and antihydrogen would provide stringent tests of the CPT (charge conjugation/parity transformation/time reversal) invariance and the weak equivalence principle [1]. In the last few years, the ALPHA collaboration has produced [2], and trapped antihydrogen [3, 4]. Recently, ALPHA studied antihydrogen's internal structure by inducing hyperfine transitions in ground state atoms [5].

In order to form antihydrogen, antiprotons and positrons are first stored in the form of non-neutral plasmas in Penning-Malmberg traps [6] and then, are allowed to interact to form antihydrogen [7].

The radial compression of antiproton, electron and positron plasmas is necessary to counteract expansion drag due to asymmetries in the static fields and the presence of background gases [8–10], and thereby attain long confinement times. Moreover, radial compression allows control of the radial sizes and densities of the non-neutral plasmas [11]. A commonly-used technique is the rotating wall (RW), in which a time-varying azimuthal rotating electric field is used to balance or exceed the drag by applying a positive torque to the plasma (see e.g., [6, 12]).

Antiproton cloud compression is an important tool for the formation and trapping of cold antihydrogen. Decreasing the antiproton cloud's radius reduces the circumferential velocity of the antiprotons and results in antihydrogen atoms with lower kinetic energy [13]. Additionally, ALPHA's magnetic trap is used to confine low energy antihydrogen atoms and it is composed of an octupole magnet providing a transverse magnetic field, plus two mirror coils [14]. The transverse magnetic field breaks the cylindrical symmetry of the

8 Physics Department, CERN, CH-1211 Geneva 23, Switzerland

9 Department of Physics and Astronomy, York University, Toronto, Ontario, M3J 1P3, Canada

10 Instituto de Física, Universidade Federal do Rio de Janeiro, Rio de Janeiro 21941-972, Brazil

11 Department of Physics and Astronomy, Aarhus University, DK-8000 Aarhus C, Denmark

12 TRIUMF, 4004 Wesbrook Mall, Vancouver, British Columbia V6T 2A3, Canada

13 Department of Physics, Stockholm University, SE-10691 Stockholm, Sweden

14 Department of Physics, University of Liverpool, Liverpool L69 7ZE, UK

15 Department of Physics, Purdue University, West Lafayette, Indiana 47907, USA

16 Department of Physics, NRCN-Nuclear Research Center Negev, Beer Sheva, IL-84190, Israel

17 Department of Physics and Astronomy, University of Calgary, Calgary, Alberta T2N 1N4, Canada

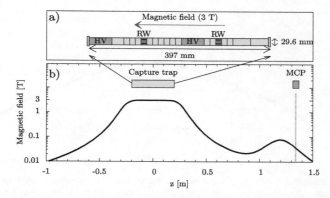

Fig. 1 a) Schematic of the electrodes making up the Penning-Malmberg trap of the antiproton capture trap. b) The magnetic field on the axis of the trap as a function of the longitudinal position. The position of the electrodes and the MCP/phosphor/CCD detector are illustrated. A small solenoid is placed at $z = 1.2$ m to guide the particles

Penning-Malmberg trap and induces non-neutral plasma diffusion [15] and ballistic loss [16]. The exposure of the plasmas to the octupole's transverse magnetic field can be minimized by reducing their radial size. Finally, the antiproton and positron plasmas should be well-overlapped to maximize the yield of antihydrogen atoms. Since radially small, dense positron plasmas are needed to increase the rate of antihydrogen production, thus radially small antiproton clouds are optimal.

Until present, two different kinds of antiproton cloud compression have been reported. In Ref. [17], in the ALPHA experiment, an electron plasma co-located in the trap was compressed by a RW in the 10 MHz region, cooling and sympathetically compressing the antiproton cloud. In this work, we present evidence of compression of antiproton clouds at low frequencies (hundreds of kHz), in a markedly different regime. Furthermore, this work is differentiated from the observations reported by the ASACUSA collaboration in Ref. [18], since we use an electron plasma as a source of cooling for the antiprotons. The presence of the electron plasma, particularly its self-electric field, greatly affects the behaviour of the system.

2 Experimental procedure

2.1 Apparatus

The upgraded ALPHA antiproton capture trap used for these experiments is a Penning-Malmberg trap with a stack of twenty cylindrical electrodes for axial confinement of charged particles, plus a 3 T solenoidal magnetic field, directed along the trap axis, to confine the charged particles radially. Figure 1a illustrates the electrode stack. Two high-voltage (HV) electrodes are used to catch and trap antiprotons from the Antiproton Decelerator [14].

The particles can be released from the trap onto a MCP/phosphor/CCD[1] detector assembly to destructively image the radial density profile [19]. The detector is shown on the right hand side of Fig. 1b, along with a plot of the axial magnetic field used to guide the particles.

---

[1] MCP: micro-channel plate and CCD: charge-coupled device.

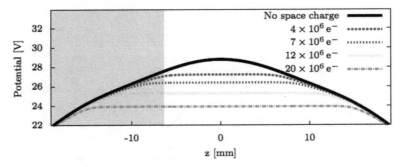

Fig. 2 Black continuous line: the electric potential well used to hold the electrons during the application of the RW. The (purple) shaded region indicates the position of the rotating wall electrode. The various dotted and dashed lines show the self-consistent potential for the numbers of electrons indicated in the legend

2.2 Antiproton capture and cooling using secondary electrons

Antiprotons are extracted from the Antiproton Decelerator into the experiment with a kinetic energy of 5.3 MeV. The energy of the antiprotons is degraded by thin layers of aluminium and beryllium, and antiprotons with an energy less than 5 keV are captured by the high-voltage electrodes [14]. The antiprotons are cooled by allowing them to interact with an electron plasma [20]. Energy is transferred to the electrons through Coulomb collisions, while the electrons cool with a time constant of about 0.4 s in the 3 T magnetic field through emission of cyclotron radiation.

Typically, the electrons are preloaded from a source, but in this work, we made use of the secondary electrons that are created when the antiprotons pass through the degrader layers. Using secondary electrons, ~ 90 % of the antiprotons are cooled while usually only ~ 60 % are cooled with preloaded electrons [17]. This increase in the cooling efficiency is due to improved radial overlap of the antiproton cloud and the secondary electrons. For every measurement, this cooling procedure results in $\sim 1.5 \times 10^5$ antiprotons and $\sim 20 \times 10^6$ electrons. If desired, a fraction of the electrons can be removed by suddenly opening one side of the trap well. Depending on the pulse time and voltage, a fraction of the electrons escape from the trap, while heavier antiprotons remain trapped [14].

2.3 Rotating wall application

The RW field is produced by an electrode divided into azimuthally isolated segments and by applying to each segment a sinusoidal potential $V_j(t)$ of frequency ω, amplitude A and phase $\theta_j = 2\pi j/k$, where k is the number of segments. The potential can be expressed as:

$$V_j(t) = A\cos(\theta_j - \omega t). \tag{1}$$

For the measurements presented in this paper, we used one of two six-segment electrodes (identified as RW in Fig. 1). For each measurement, antiprotons and electrons were captured and cooled. Then, the electron number was adjusted if necessary and the RW was applied at a fixed amplitude of 1 V for 100 s. For technical reasons, the frequency of the drive was swept over a 0.2 kHz range centred on a given frequency. After the RW application, the particles were extracted onto the MCP.

The continuous line in Fig. 2 shows the potential well used to hold the particles while the RW was applied. The potential well is almost harmonic with an antiproton bounce frequency

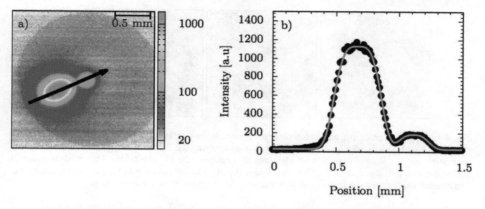

Fig. 3 a) MCP image of an antiproton-electron plasma after compression. The (black) arrow indicates the position of the profile shown in b). b) Dots are the data from the radial profile across the arrow shown in a) and the (red) curve is the respective fit

of ~ 270 kHz. When electrons are added, the shape of the potential is distorted due to their space charge. The dotted and dashed lines in Fig. 2 shows the total self-consistent potential when using 4×10^6, 7×10^6, 12×10^6 and 20×10^6 electrons, as calculated by solving Poisson's equation with a density distribution given by a Boltzmann distribution, using the self-consistent potential [6].

2.4 Analysis of MCP images

An example image of an antiproton-electron plasma is shown in Fig. 3a. Due to their mass difference, the antiprotons and electrons image to different positions on the MCP, with the antiprotons appearing on the left [19, 21]. The electron density can be conveniently described by a two-dimensional generalized Gaussian of the form $n_e \exp(|\frac{\mathbf{r}-\mathbf{r}_e}{\sigma_e}|^{k_e})$, where n_e, σ_e, $\mathbf{r}_e$ and k_e are fit parameters. For the antiproton density, we use a similar equation but modified to account for the observed elliptical shape [21]. A simultaneous fit of the two distributions is performed for each image. A cut through the image in Fig. 3a is plotted in Fig. 3b, along with the respective fit. n_e and $n_{\overline{p}}$, the central densities of the electrons and the antiprotons, respectively, are used as quantitative measures of the degree of compression.

3 Results

An image of the antiproton-electron plasma after capture and cooling is shown in Fig. 4a. We estimate that the plasma has a radius of ~ 4 mm and a rotation frequency of ~ 10 kHz. Figure 4b shows the plasma after the electron number has been reduced to 4×10^6 electrons and a RW field at a frequency of 140 kHz has been applied for 100 s. We observe that the antiproton cloud has been compressed to a radius of ~ 0.3 mm, while only a few of the electrons have been compressed. Figure 4c shows the plasma with 20×10^6 electrons after compression at 600 kHz. Similarly to the previous case, we see a dense antiproton cloud, with a radius of ~ 0.2 mm. Additionally, about 15 % of the electron plasma has also been compressed. In any of these experiments, we do not observe that the rotating wall induces any loss of antiprotons.

 ⚛ Springer

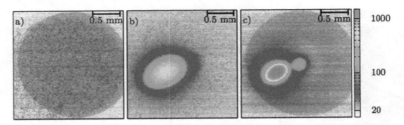

Fig. 4 a) MCP image of the antiproton-electron plasma before applying the RW. b) MCP image of the antiproton cloud co-located with 4×10^6 electrons after applying the RW at 140 kHz. c) MCP image of the antiproton cloud co-located with 20×10^6 electrons after applying the RW at 600 kHz. The clouds in image a) are too disperse to image as b) and c). The circle near the edges of images a) and c) is a mechanical aperture. The mechanical aperture is not observed in b) because the electrons are not dense enough

The striking difference that emerges when compared to the case of sympathetic compression [17] is that when the RW field is applied to a pure electron plasma in the hundreds of kHz range, no compression is observed. (Recall in [17] that the electron plasma was compressed with or without the presence of the antiprotons). This implies that in the present case, the antiproton compression was caused by the RW field directly coupling to the antiprotons, rather than being mediated by compression of the electron plasma. On the other hand, when no electrons are present and the RW is applied to a pure antiproton cloud, no compression is observed. Antiprotons are lost from the trap, perhaps indicating that they are heated by the RW field in the absence of a cooling medium.

The compression was studied as a function of RW frequency for antiproton-electron plasmas containing different numbers of electrons. Figure 5 shows the antiproton cloud central density, $n_{\bar{p}}$ (see Section 2.4), as a function of the RW frequency for different numbers of electrons.

One can see, for the 4×10^6 electrons case, that the antiproton cloud compresses for RW frequencies in the range 50–200 kHz. With increasing numbers of electrons, the maximum frequency at which compression is observed increases. Moreover, higher central densities are achieved with larger numbers of electrons. We note that the antiproton cloud does not compress well above ~ 700 kHz, which is similar to the lowest frequency at which the pure electron plasma compresses.

4 Bounce resonant transport of antiprotons

In many experiments, the RW field couples to Trivelpiece-Gould (TG) modes of the plasma, thereby applying a torque and leading to radial compression of the plasma [22, 23]. However, the lowest TG mode frequency of the electron plasma studied here is ~ 15 MHz, and therefore this mechanism is not consistent with the observed compression.

At first glance, compression by magnetron sideband cooling seems to provide an explanation for the data, because the sum of the magnetron (~ 1 kHz) and axial bounce (~ 270 kHz) frequencies of the antiprotons is close to the compression frequency [24–27]. Magnetron sideband cooling requires harmonic potentials, although compression has recently been achieved for independent particles in a slightly anharmonic potential [28]. This mechanism is unlikely to be responsible for the observed compression, because the space charge of the electrons greatly distorts the potential and it becomes highly anharmonic.

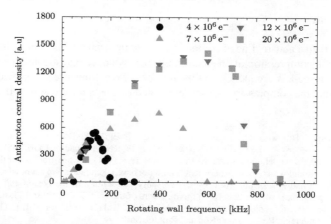

Fig. 5 Antiproton central density after applying the RW for 100 s, at 1 V and at a chosen frequency (with 0.2 kHz sweep). Different number of electrons are used, while the antiproton number remains the same at $\sim 1.5 \times 10^5$. The error bars are too small to be visible

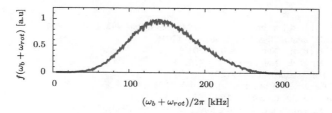

Fig. 6 The distribution of $(\omega_b + \omega_{rot})$ for antiprotons cooled by 4×10^6 electrons

For particles moving in an asymmetric time-varying potential (such as a RW), it has been observed that resonances between the particle's motional frequency and the drive frequency can result in radial inward or outward movement [29–33]. We have investigated whether this mechanism can be responsible for the compression described in Section 3. The antiproton bounce frequency, ω_b, was calculated by integrating the one-dimensional equations of motion in the self-consistent electric potential (see Fig. 2). Taking into account the distributions of radial positions and thermal energies allows a distribution of the antiproton bounce frequencies to be built up. The rotation frequency, ω_{rot}, of the antiproton-electron plasma was calculated from the self-consistent electric potential, and is dominated by the density of the electron plasma.

Figure 6 shows the combined distribution, $f(\omega_b + \omega_{rot})$, of the antiprotons when cooling with 4×10^6 electrons. This is the simplest system, since the electron plasma can be assumed to remain constant during the application of the RW. This assumption is supported by the fact that a compressed electron plasma is not visible in Fig. 4b. We observe that $f(\omega_b + \omega_{rot})$ lies over the same range of frequencies (50 – 200 kHz) as the observed compression. This indicates that bounce resonant transport may be a viable explanation for our data. When using a larger number of electrons, the system becomes more complex. The electron plasma compresses over time, with the result that both the rotation frequency and the shape of the potential well change dynamically. Compression of the electron plasma would increase ω_{rot} and consequently shift $f(\omega_b + \omega_{rot})$ to higher frequencies. This is qualitatively consistent with our observations, but further work is needed before a definitive conclusion can be made.

5 Conclusion

We have observed a new regime of compression of antiproton clouds, where the effect is not mediated by the compression of an electron plasma. We have investigated the compression as a function of the RW frequency and the number of electrons used to provide cooling. It is suggested that the compression may be explained by the bounce resonant transport of the antiprotons.

Acknowledgments This work was supported by: CNPq, FINEP/RENAFAE (Brazil); ISF (Israel); FNU (Denmark); VR (Sweden); NSERC, NRC/TRIUMF, AITF, FQRNT (Canada); DOE, NSF, LBNL-LDRD (USA); and EPSRC, the Royal Society and the Leverhulme Trust (UK). We are grateful for the efforts of the CERN AD team, without which these experiments could not have taken place. We acknowledge the valuable work of J. Strachan and P. Morrall, from STFC Daresbury Laboratory, in the design and construction of the antiproton capture trap. We are grateful to the members of the UCSD Non-neutral Plasma and Positron Research teams for the helpful discussions, in particular to Prof. T.M. O'Neil, Prof. D.H.E. Dubin, Prof. C.F. Driscoll, Prof. F. Anderegg, Prof. J. Danielson and Prof. C. Surko.

References

1. Holzscheiter, M.H., Charlton, M., Nieto, M.M.: Phys. Rep. **402**, 1–101 (2004)
2. Andresen, G.B., et al.: Phys. Lett. B **685**, 141 (2010)
3. Andresen, G.B., et al.: Nature **468**, 673 (2010)
4. Andresen, G.B., et al.: Nat. Phys. **7**, 558 (2011)
5. Amole, C., et al.: Nature **483**, 439 (2012)
6. Dubin, D.H.E., O'Neil, T.M.: Rev. Mod. Phys. **71**, 87 (1999)
7. Amoretti, M., et al.: Nature **419**, 456 (2002)
8. Malmberg, J.H., Driscoll, C.F.: Phys. Rev. Lett. **44**, 654 (1980)
9. Eggleston, D.L., O'Neil, T.M., Malmberg, J.H.: Phys. Rev. Lett. **53**, 982 (1984)
10. Notte, J., Fajans, J.: Phys. Plasmas **1**, 1123 (1994)
11. Huang, X.-P., et al.: Phys. Plasmas **5**, 1656 (1998)
12. Huang, X.-P., et al.: Phys. Rev. Lett. **78**, 875 (1997)
13. Jonsell, S., et al.: J. Phys. B: At. Mol. Opt. Phys. **42**, 215002 (2009)
14. Amole, C., et al.: Nucl. Instr. and Meth. A **735**, 319 (2014)
15. Gilson, E.P., Fajans, J.: Phys. Rev. Lett. **90**, 015001 (2003)
16. Fajans, J., et al.: Phys. Plasmas **15**, 032108 (2008)
17. Andresen, G.B., et al.: Phys. Rev. Lett. **100**, 203401 (2008)
18. Kuroda, N., et al.: Phys. Rev. Lett. **100**, 203402 (2008)
19. Andresen, G.B., et al.: Rev. Sci. Inst. **80**, 123701 (2009)
20. Gabrielse, G., et al.: Phys. Rev. Lett. **63**, 1360 (1989)
21. Andresen, G.B., et al.: Phys. Rev. Lett. **106**, 145001 (2011)
22. Trivelpiece, A.W., Gould, R.W.: J. App. Phys. **30**, 1784 (1959)
23. Anderegg, F., et al.: Phys. Rev. Lett. **81**, 4875 (1998)
24. Wineland, D., Dehmelt, H.: Int. J. Mass Spectrom. Ion Phys. **16**, 338 (1975)
25. Brown, L.S., Gabrielse, G.: Rev. Mod. Phys. **58**, 233 (1986)
26. Kellerbauer, A., et al.: Phys. Rev. A **73**, 062508 (2006)
27. Isaac, C.A., et al.: Phys. Rev. Lett. **107**, 033201 (2011)
28. Deller, A., et al.: New. J. Phys. **16**, 073028 (2014)
29. Eggleston, D.L., O'Neil, T.M.: Phys. Plasmas **6**, 2699 (1999)
30. Greaves, R.G., Surko, C.M.: Phys. Plasmas **8**, 1879 (2001)
31. Eggleston, D.L., Carrillo, B.: Phys. Plasmas **9**, 786 (2002)
32. Eggleston, D.L., Carrillo, B.: Phys. Plasmas **10**, 1308 (2003)
33. Greaves, R.G., Moxom, J.M.: Phys. Plasmas **15**, 072304 (2008)

Hyperfine Interact (2015) 235:29–36
DOI 10.1007/s10751-015-1186-0

Narrowband solid state vuv coherent source for laser cooling of antihydrogen

J. Mario Michan[1] · Gene Polovy[2] · Kirk W. Madison[2] ·
Makoto C. Fujiwara[1] · Takamasa Momose[3]

Published online: 13 May 2015

Abstract We describe the design and performance of a solid-state pulsed source of narrowband ($<$ 100 MHz) Lyman-α radiation designed for the purpose of laser cooling magnetically trapped antihydrogen. Our source utilizes an injection seeded Ti:Sapphire amplifier cavity to generate intense radiation at 729.4 nm, which is then sent through a frequency doubling stage and a frequency tripling stage to generate 121.56 nm light. Although the pulse energy at 121.56 nm is currently limited to 12 nJ with a repetition rate of 10 Hz, we expect to obtain greater than 0.1 μJ per pulse at 10 Hz by further optimizing the alignment of the pulse amplifier and the efficiency of the frequency tripling stage. Such a power will be sufficient for cooling a trapped antihydrogen atom from 500 mK to 20mK.

Keywords Lyman-Alpha · VUV · Antihydrogen · Laser Cooling

1 Introduction

As the simplest and best understood atom in the periodic table, hydrogen appeared to be the natural choice for early laser cooling experiments. Despite this, the first and only study

Proceedings of the 6th International Conference on Trapped Charged Particles and Fundamental Physics (TCP 2014), Takamatsu, Japan, 1-5 December 2014

✉ Takamasa Momose
   momose@chem.ubc.ca

1   TRIUMF, 4004 Wesbrook Mall, Vancouver, BC V6T 2A3, Canada

2   Department of Physics and Astronomy, The University of British Columbia,
   6224 Agricultural Road, Vancouver, BC V6T 1Z1, Canada

3   Department of Chemistry, Department of Physics and Astronomy,
   The University of British Columbia, 2036 Main Mall,
   Vancouver, BC V6T 1Z1, Canada

 Springer

of the optical cooling of hydrogen was published in 1993 [16], many years after several alkalis were successfully cooled and trapped. The reason for this is the inherent difficulty of producing coherent radiation at the laser cooling transition for hydrogen - 121.56 nm or Lyman-α. Numerous broadband pulsed Lyman-α sources were developed in the late 1970s and early 1980s [2, 4, 9–13, 17]. These were suitable for many spectroscopic applications, but not for laser cooling, which requires, for efficient cooling, the linewidth of the source to be less than or equal to the natural linewidth ($\Gamma \approx 100$ MHz) of the transition [16]. The first narrowband ($\Delta \nu \approx 40$ MHz) Lyman-α source was developed in 1987 [3] based on non-resonant third-harmonic generation of frequency-doubled pulse-amplified light from a tunable CW dye laser, which made it possible to laser cool hydrogen, trapped in a magnetic trap, to less than 8 mK [16]. Nevertheless, low repetition rate pulsed sources cannot be used for conventional laser cooling experiments, which rely on optical molases, and interest in coherent Lyman-α sources subsided until a new application emerged - laser cooling of trapped antihydrogen. In particular, the ALPHA collaboration at CERN successfully trapped antihydrogen in a magnetic trap at a translational temperature of 500 mK and plans to use the trapped antimatter to test for charge, parity and time (CPT) symmetry violations and to investigate matter-antimatter gravity interactions [1, 5]. To reach the level of accuracy required for these measurements, the trapped antihydrogen must be cooled to a translation temperature of ~ 20 mK. According to simulations [5], a high power (~ 1 μW) Lyman-α source with a 100 MHz linewidth is required to perform the antihydrogen cooling. Two continuous wave (CW) Lyman-α sources, which rely on a four-wave mixing scheme in mercury vapour, have been realized [8, 15], but they are not yet capable of delivering sufficient power (and one of them relies on dye lasers). Here, we present an alternative approach - a narrowband (< 100 MHz) pulsed solid-state Lyman-α source. Our motivation was to develop a completely solid-state source to avoid the maintenance and additional safety issues associated with a dye-laser-based source. Simulations have shown that a narrowband pulsed source that produces 0.1 μJ per shot at a repetition rate of 10 Hz is capable of cooling a trapped antihydrogen atom down to 20 mK [5]. The goal of our development is to achieve this level of 121.56 nm pulse energy with a solid-state laser system.

2 Apparatus

The laser system shown in Fig. 1 consists of a pulsed Ti:Sapphire amplifier seeded by a narrow linewidth (< 1 MHz) semiconductor laser (Toptica Photonics TA Pro) operating at 729.4 nm, an anti-reflection coated type I BBO crystal (Castech Inc.) for second harmonic generation (SHG) at 729.4 nm and a third harmonic generation (THG) and detection chamber.

2.1 Ti: sapphire amplifier and SHG

To obtain the intensities required for efficient SHG and THG while maintaining a narrow linewidth, we used an injection seeded pulsed Ti:Sapphire amplifier based on [7]. This amplifier is an unstable resonator that consists of two Brewster cut Ti:Sapphire crystals (10 mm in diameter, 36.8 mm in path length, GT Advanced Technologies), a convex graded reflectivity mirror (f = -5 m, INO) coated for 729.4 nm, a high reflector (HR) and two isosceles Brewster prisms for coarse wavelength selection. As demonstrated previously in [7], the generation of the Fourier-limit pulses can be achieved when each crystal is pumped

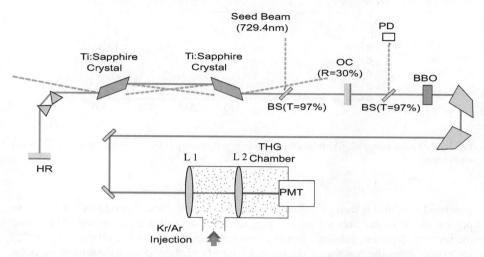

Fig. 1 Laser arrangement and non-linear optical stages. The Ti:Sapphire amplifier cavity is seeded by a narrow linewidth semiconductor laser via an uncoated CaF_2 beam splitter (BS) with a reflectivity of $\sim 3\%$. The two Ti:Sapphire crystals are pumped from both sides by a pulsed Nd:YAG laser. The pump power is distributed evenly between the four arms. The amplified light, which exits the cavity via the output coupler (OC), is then sent through a coated BBO crystal. The second harmonic is separated from the fundamental beam (not shown), with two Pellin Broca prisms and routed to the THG chamber. The beam is then focused by an MgF_2 lens L_1 (f = 100 mm) onto a mixture of krypton and argon gas. Finally, Lyman-α radiation is produced by non-resonant THG, recollimated by an MgF_2 lens L_2 (f = 200 mm) and detected by a solar-blind photo-multiplier tube (PMT). Two Lyman-α filters are placed between L_2 and the PMT to prevent 364.7 nm light from reaching the PMT (not shown)

by a pulsed frequency doubled Nd:YAG laser (532 nm, 10 Hz repetition rate) from both sides and injection seeded with a CW diode laser ($\Delta \nu < 1$ MHz, 729.4 nm). The output of the amplifier is then sent into a coated type I BBO crystal (dimensions: 7 mm $\times$ 7 mm $\times$ 8 mm), to produce 364.7 nm radiation. After the second harmonic (364.7 nm) is spatially separated from the fundamental beam (729.4 nm) by means of two Pellin Broca prisms, it is sent to the THG chamber, where we produce and detect Lyman-α radiation.

2.2 Lyman-α generation and detection

To generate Lyman-α radiation, we used non-resonant THG in a mixture of krypton and argon - an approach that has been demonstrated a number of times in the literature [2–4, 9, 10, 12, 17]. A resonant four-wave-mixing (FWM) scheme was also successfully demonstrated with a broadband dye-laser source by Michan et al. [14]. However, THG (which requires only one color) was chosen here because of its technical simplicity and the availability of high energy pump sources. In Fig. 1, we show that radiation from the SHG stage is sent into the THG chamber, which consists of an MgF_2 focusing lens L_1 (f = 100 mm), an input port for the krypton-argon mixture, an MgF_2 recollimating lens L_2 (f = 200 mm), two Lyman-α filters (Pelham Research Optical L.L.C. 122-NB) and a solar-blind photo-multiplier tube (PMT: Hamamatsu R972). In our implementation, the first lens L_1 focuses the slowly diverging 364.7 nm beam (with a vertical diameter of 6 mm and a horizontal diameter of 3 mm at 3 m away from the BBO crystal, based on a burn card measurement) onto the gas mixture to reach the intensities required for efficient THG. The

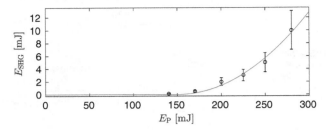

Fig. 2 SHG pulse energy at 365 nm as a function of the pulse energy of the 532 nm pump for the 729.4 nm seeded amplifier

generated Lyman-α is then recollimated with lens L_2, sent through two Lyman-α filters (to prevent the 364.7 nm pulses from reaching the detector) and detected by the PMT. Using the response function, gain curve and quantum efficiency of the PMT (Hamamatsu R972) we extract from the raw signal the temporal profile of the Lyman-α beam and its pulse energy. To confirm the production of Lyman-α, we verified that the PMT signal vanished completely in the absence of krypton in the THG chamber.

3 Results

In this section, we describe the results we obtained at the SHG and THG stages. A detailed description and characterization of the injection seeded Ti:Sapphire amplifier can be found in [7] and [6].

3.1 Second harmonic generation stage

3.1.1 Pulse energy

In Fig. 2, the average pulse energy of the second harmonic (364.7 nm) E_{SHG} is plotted as a function of the 532 nm pump pulse energy E_P. This data was taken with the cavity of the Ti:Sapphire amplifier unlocked (free run), and therefore large fluctuations in the pulse-to-pulse energy were observed due to the fluctuation of the cavity length. We fit this data to the following model:

$$E_{\text{SHG}} = \alpha \cdot (E_P - E_{\text{TH}})^2 \cdot \Theta (E_P - E_{\text{TH}}) \tag{1}$$

where $\alpha = (5.7 \pm 2.8) \times 10^{-4} \text{mJ}^{-1}$ is a proportionality constant, $E_{\text{TH}} = 150 \pm 27$ mJ is the pump power at the 729.4 nm lasing threshold and Θ is the Heaviside step function.

Further optimizations of the alignments of the amplifier cavity and pump geometries enabled us to reach the SHG pulse energies shown in Fig. 5 with $E_P = 310$ mJ, where the mean pulse energy is 16.0 mJ and the standard deviation is 1.4 mJ. In this example, the amplifier cavity was locked to off-resonance with the seed frequency, in which the transmittance of the seed laser radiation from the cavity was near the minimum. A slow fluctuation (~ 20 sec) in the pulse energy seen in Fig. 5 might be due to the temperature fluctuation of the BBO crystal, which will be removed in the near future.

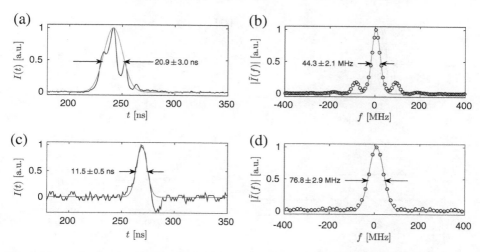

Fig. 3 Temporal and spectral profiles of the 364.7 nm and 121.56 nm pulses with the amplifier cavity locked to near-resonance with the seed laser. **a** Temporal profile of a single pulse at 364.7 nm. Black: Observed intensity. Red: Fitted curve with a Gaussian function. **b** Spectral profile of the 364.7 nm pulse. Black: Fourier transform of the black trace in **a**. Red: Fitted curve with a Gaussian function. **c** Temporal profile of the 121.56 nm pulse. Black: Observed intensity. Red: Fitted curve with a Gaussian function. **d** Spectral profile of the 121.56 nm pulse. Black: Fourier transform of the black trace in **a**. Red: Fitted curve with a Gaussian function

3.1.2 Pulse shape

Figure 3a and b, respectively, shows an instantaneous intensity (arbitrary units) of a 364.7 nm pulse as a function of time and the corresponding spectrum when the amplifier cavity was near-resonance with the seed frequency, in which the transmittance of the seed laser radiation from the cavity was near the maximum. The spectrum (b), which is obtained by taking a Fourier transform of the temporal profile shown in (a), consists of a dominant peak (which corresponds to the envelope of the temporal profile) and two side lobes, which account for the modulation seen in the temporal profile. Since a simple side-locking technique was employed for the cavity locking, there were still a non-negligible detuning of the cavity length with respect to the seed laser wavelength, which resulted in the excitation of the adjacent cavity mode via the mode coupling [7]. The frequency separation between the dominant peak and the side lobe is about 95 MHz, which roughly corresponds to the free spectral range (FSR) of the cavity (length ~ 1.55 m). The full width at half maximum (FWHM) of the temporal profile Δt_{FWHM} and the FWHM of the spectrum $\Delta \nu_{FWHM}$ are 20.9 ± 3.0 ns and 44.3 ± 2.1 MHz, respectively. Similarly, the temporal profile and spectrum of a pulse generated when the amplifier cavity is off-resonance with the seed frequency are shown in Fig. 4a and b. When the amplifier cavity was off-resonance, the side-bands were more enhanced, which account for the deeper modulation seen in the temporal profile. For this pulse, $\Delta t_{FWHM} = 21.3 \pm 2.0$ ns and $\Delta \nu_{FWHM} = 42.7 \pm 1.4$ MHz. The pulse energy at 364.7 nm is 5.0 mJ and 7.2 mJ at 250 mJ pump energy (532 nm) for the near-resonant (Fig. 3) and off-resonant (Fig. 4) conditions, respectively. The off-resonant cavity condition results in higher total pulse energy than the near-resonant condition due to the stronger mode coupling, which was also reported in [7].

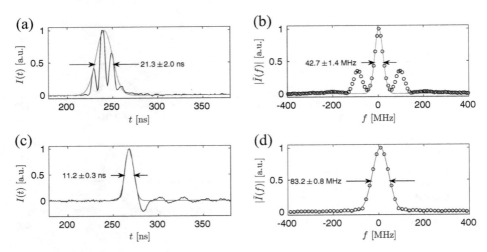

Fig. 4 Temporal and spectral profiles of the 364.7 nm (**a** and **b**) and 121.56 nm (**c** and **d**) pulses with the amplifier cavity locked to off-resonance with the seed laser. See the caption of Fig. 3

3.2 THG stage

The Lyman-α power is proportional to $\chi^2 N^2 I_{SHG}^3 F$, where χ is the third order susceptibility, N is the number density of the krypton, I_{SHG} is the power of the 364.7 nm beam and F is the phase matching factor. In the limit of a tight focus, the phase-matching factor is maximized when $b \cdot \Delta k = b \cdot (k_{121.56 \text{ nm}} - 3 \cdot k_{364.7 \text{ nm}}) = -2$, where b is the confocal parameter of the beam (assuming a TEM$_{00}$ mode) [9]. However, because the Lyman-α power is also proportional to N^2, adjusting the krypton density to satisfy the phase matching condition alone is insufficient to optimizing the Lyman-α power and the problem becomes a three parameter optimization for the focal length of the focusing lens L_1 and the pressures of krypton and argon in the mixing chamber. Based on previous results from preliminary dye-laser experiments and some empirical optimization, we chose an f = 100 mm MgF$_2$ lens for L_1 and set the krypton pressure to 84 mTorr and the argon pressure to 200 mTorr for the present study.

With the amplifier cavity locked to the near-resonance, we observed the temporal profile for the 121.6 nm pulse as shown in Fig. 3c. The corresponding spectrum shown in Fig. 3d is obtained by the Fourier transform of the temporal profile in (c). The generated Lyman-α pulse has a temporal width of $\Delta t_{FWHM} = 11.5 \pm 0.5$ ns and a spectral width of 76.8 ± 2.9 MHz. Note that the temporal width of the 121.56 nm pulse in (c) might be slightly broadened due to the response time of the PMT (the anode pulse rise time of 1.6 ns and the electron transit time of 17 ns). As a result, the spectral width obtained in (d) might be slightly narrower than the actual spectral width. Corresponding temporal profile and spectrum for the 121.6nm pulse with the off-resonant cavity condition are shown in Fig 4c and d, respectively. The generated 121.56 nm pulse has similar temporal and spectral profiles; $\Delta t_{FWHM} = 11.2 \pm 0.3$ ns and $\Delta \nu = 83.2 \pm 0.8$ MHz. The spectral width is slightly broadened with this condition.

Remarkably, we found that locking the cavity in the off-resonant condition with the seed laser resulted in higher THG efficiency than the cavity in the near-resonant condition. At 250 mJ of 532 nm pump energy, the off-resonant condition shown in Fig. 4

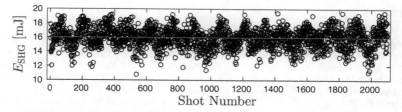

Fig. 5 SHG pulse energy with $E_P = 310$ mJ as a function of time with the amplifier cavity locked. The mean pulse energy is 16.0 mJ and the standard deviation is 1.4 mJ

resulted in 12 nJ pulse energy for the 121.46 nm pulse. On the other hand, the near-resonant condition shown in Fig. 3 resulted in only 2 nJ pulse energy for the 121.46 nm pulse. The higher THG efficiency with the off-resonant condition is most likely due to the higher peak power density of each pulses in the pulse train at 364.7 nm shown in Fig. 4 a. For the laser cooling of antihydrogen, whose whose natural linewidth is expected to be $\Gamma \approx 100$ MHz, the off-resonant condition for the amplifier cavity would be better than the near-resonant condition because of the higher pulse energy at 121.46 nm.

4 Conclusion and outlook

Here, we have developed a solid-state pulsed source of narrowband (< 100 MHz) Lyman-α radiation. The maximum Lyman-α output we have demonstrated here - by frequency tripling a 7.2 mJ pulse of 364.7 nm light - was 12nJ. Due to the present environmental conditions (humidity and temperature) of the facility, we were forced to lower the pulse energy of 532 nm in order to avoid any optical damage on the Ti:Sapphire crystals. For this reason, the data presented in Section 3.2 were taken at the 532 nm pump energy of 250 mJ. Since we have observed about 20mJ of 364.7 nm radiation as the maximum energy of a single shot at the 532 nm pump energy of 310 mJ (Figs. 2 and 5), we are confident that, after alleviating problems due to the environment, more than 0.1 μJ is possible with this system. The pulse energy at 121.5 nm generated by THG is proportional to the cube of the pulse energy at 364.7 nm, and therefore, if the conversion efficiency remains constant, simply increasing the SHG pulse energy to 20 mJ would result in approximately 0.26 μJ of Lyman-α. In addition, further improvement of the THG conversion efficiency would be expected by optimizing the krypton and argon pressures in the THG chamber and the focal length of the focusing lens (L_1) in order to achieve better phase matching condition for the THG process. While we have made significant progress toward developing a narrowband solid state Lyman-α source, further optimization of the present system will be required to achieve sufficient pulse energies with stable operation in order to use this Lyman-α source for laser cooling trapped antihydrogen, from 500 mK to 20 mK. Further development is underway.

Acknowledgments We acknowledge Prof. Terry Miller at Ohio State University and his group members for providing us technical details of their pulse amplifier system. The present work was supported by a National Science and Engineering Research Subatomic Physics Project Grant in Canada and funds from Canada Foundation for Innovation for the Centre for Research on Ultra-Cold Systems (CRUCS) at UBC.

References

1. Andresen, G., Ashkezari, M., Baquero-Ruiz, M., Bertsche, W., Bowe, P.D., Butler, E., Cesar, C., Chapman, S., Charlton, M., Deller, A., et al.: Trapped antihydrogen. Nature **468**(7324), 673–676 (2010)
2. Batishche, S., Burakov, V., Kostenich, Y., Mostovnikov, V., Naumenkov, P., Tarasenko, N., Gladushchak, V., Moshkalev, S., Razdobarin, G., Semenov, V., Shreider, E.: Optimal conditions for third-harmonic generation in gas mixtures. Opt. Commun. **38**(1), 71–74 (1981)
3. Cabaret, L., Delsart, C., Blondel, C.: High resolution spectroscopy of the hydrogen lyman-line stark structure using a vuv single mode pulsed laser system. Opt. Commun. **61**(2), 116–119 (1987)
4. Cotter, D.: Tunable narrow-band coherent VUV source for the lyman-alpha region. Opt. Commun. **31**(3), 397–400 (1979)
5. Donnan, P.H., Fujiwara, M.C., Robicheaux, F.: A proposal for laser cooling antihydrogen atoms. J. Phys. B: Atomic Mol. Opt. Phys. **46**(2), 025–302 (2013)
6. Dupré, P.: Modeling a nanosecond quasi-fourier-transform limited ti: Sa laser source. Eur. Phys. J. Appl. Phys. **40**(03), 275–291 (2007)
7. Dupré, P., Miller, T.A.: Quasi-fourier-transform limited, scannable, high energy titanium-sapphire laser source for high resolution spectroscopy. Rev. Sci. Instrum. **78**(3), 033102 (2007)
8. Eikema, K.S.E., Walz, J., Hänsch, T.W.: Continuous wave coherent lyman-α radiation. Phys. Rev. Lett. **83**, 3828–3831 (1999)
9. Hilbig, R., Wallenstein, R.: Enhanced production of tunable vuv radiation by phase-matched frequency tripling in krypton and xenon. IEEE J. Quantum Electron. **17**(8), 1566–1573 (1981)
10. Langer, H., Puell, H., Röhr, H.: Lyman alpha (1216 å) generation in krypton. Opt. Commun. **34**(1), 137–142 (1980)
11. Mahon, R., Tomkins, F.S., Kelleher, D.E., McIlrath, T.J.: Four-wave sum mixing in beryllium around hydrogen lyman-α. Opt. Lett. **4**(11), 360–362 (1979)
12. Mahon, R., Yiu, Y.M.: Generation of lyman-α radiation in phase-matched rare-gas mixtures. Opt. Lett. **5**(7), 279–281 (1980)
13. McKee, T.J., Stoicheff, B.P., Wallace, S.C.: Tunable, coherent radiation in the lyman-α region (1210–1290 å) using magnesium vapor. Opt. Lett. **3**(6), 207–208 (1978)
14. Michan, J.M., Fujiwara, M.C., Momose, T.: Development of a lyman-laser system for spectroscopy and laser cooling of antihydrogen. Hyperfine Interact. **228**(1-3), 77–80 (2014)
15. Scheid, M., Kolbe, D., Markert, F., Hänsch, T.W., Walz, J.: Continuous-wave lyman-α generation with solid-state lasers. Opt. Express **17**(14), 11,274–11,280 (2009)
16. Setija, I.D., Werij, H.G.C., Luiten, O.J., Reynolds, M.W., Hijmans, T.W., Walraven, J.T.M.: Optical cooling of atomic hydrogen in a magnetic trap. Phys. Rev. Lett. **70**, 2257–2260 (1993)
17. Wallenstein, R.: Generation of narrowband tunable VUV radiation at the lyman-wavelength. Opt. Commun. **33**(1), 119–122 (1980)

Hyperfine Interact (2015) 235:37–44
DOI 10.1007/s10751-015-1199-8

Status of deceleration and laser spectroscopy of highly charged ions at HITRAP

Zoran Andelkovic[1] · Gerhard Birkl[2] · Svetlana Fedotova[1] · Volker Hannen[3] ·
Frank Herfurth[1] · Kristian König[2] · Nikita Kotovskiy[1] · Bernhard Maaß[2] ·
Jonas Vollbrecht[3] · Tobias Murböck[2] · Dennis Neidherr[1] ·
Wilfried Nörtershäuser[2] · Stefan Schmidt[2] · Manuel Vogel[2] ·
Gleb Vorobjev[1] · Christian Weinheimer[3]

Published online: 8 July 2015
© Springer International Publishing Switzerland 2015

Abstract Heavy few-electron ions are relatively simple systems in terms of electron
structure and offer unique opportunities to conduct experiments under extremely large electromagnetic fields that exist around their nuclei. However, the preparation of highly charged
ions (HCI) has remained the major challenge for experiments. As an extension of the existing GSI accelerator facility, the HITRAP facility was conceived as a multi-stage decelerator
for HCI produced at high velocity. It is designed to prepare bunches of around 10^5 HCI and
to deliver them at low energies to various experiments. One of these experiments is Spec-
Trap, aiming for laser spectroscopy of trapped, cold HCI. We present the latest results on
deceleration of ions in a radio-frequency quadrupole, synchrotron cooling of electrons in
a trap as a preparation step for the prospective electron cooling of the HCI decelerated in
HITRAP, as well as laser cooling of singly charged Mg ions for sympathetic cooling of HCI
in SpecTrap.

Keywords Highly charged ions · Deceleration · Penning trap · Sympathetic cooling ·
Laser spectroscopy

Proceedings of the 6th International Conference on Trapped Charged Particles and Fundamental
Physics (TCP 2014), Takamatsu, Japan, 1-5 December 2014

✉ Zoran Andelkovic
z.andelkovic@gsi.de

[1] GSI Helmholtzzentrum für Schwerionenforschung GmbH, 64291 Darmstadt, Germany

[2] Technische Universität Darmstadt, 64289 Darmstadt, Germany

[3] Universität Münster, 48149 Münster, Germany

1 Introduction

Heavy, highly charged ions (HCI) like the hydrogen-like U^{91+} or Bi^{82+} represent simple systems in terms of electron structure which offer opportunities for cutting edge tests of quantum electrodynamics (QED) in the extreme fields that exist around their nuclei [1, 2]. The HITRAP project [3] at the GSI Helmholtz Centre for Heavy Ion Research and the Facility for Antiproton and Ion Research (FAIR) was started several years ago with the goal of preparing large bunches of such ions at very low energies and distributing them to different associated experiments. The experiments include, but are not limited to, laser spectroscopy with trapped HCI [4], measurements of the bound-electron g-factor [5], study of multiple electron transfer in cold atom-HCI collisions [6] and investigation of interaction between HCI and highly intense laser light [7]. Such experiments with heavy HCI at GSI were so far hampered by the relatively large energy uncertainty of the HCI produced by the accelerator facility, resulting in e.g. large Doppler width of the transition and uncertainty of the ion velocity for laser spectroscopy experiments [8]. In experiments with HCI produced by an EBIT [9] the ion energy is significantly lower, but the experimental precision is still limited by the ion temperature or simply by the low yield of the high charge states.

At GSI/FAIR, HCI are produced by acceleration and in-flight stripping of electrons in several steps. The ions can then be stored in the experimental storage ring (ESR) for experiments at energies between 400 MeV/u and 4 MeV/u. The latter is the lowest energy at which ions can still be efficiently stored in the ESR. By taking ion bunches of some 10^6 ions from the ESR, pre-decelerated to 4 MeV/u, the HITRAP facility is designed to reduce their energy further in two linear decelerators and finally in a Penning trap all the way into the sub-eV range. The cold HCI can then be forwarded at a chosen transport energy towards different experimental setups along the beamline.

2 The HITRAP linear decelerator

The first stage of the linear decelerator consists of a double-drift buncher (DDB) and an interdigital H-type structure (IH). The DDB preconditions the beam to the longitudinal acceptance of the IH structure, which decelerates the ions from 4 MeV/u down to 500 keV/u. The second stage comprises an intermediate rebuncher (RB) and a four-rod radio frequency quadrupole (RFQ) decelerator. The RB ensures maximum efficiency when the beam is injected into the RFQ decelerator, which slows down ions from 500 keV/u to 6 keV/u. Both deceleration stages run at about 108 MHz and require a peak power of up to 200 kW and 80 kW, respectively (Fig. 1).

The IH-decelerator was successfully commissioned several years ago with deceleration efficiencies close to the theoretical maximum [10] of some 60 %. The major breakthroughs were the installation of an energy-sensitive detector and the energy reduction of the ions injected into the ESR down to 30 MeV/u for purposes of commissioning, which eliminated one deceleration step in the ring and enabled up to two ejections towards HITRAP per minute, speeding up setup and optimization.

The commissioning of the RFQ decelerator has proven to be more challenging because of the very large parameter space combined with a relatively low acceptance of the device. The sampling of the full parameter space is very time consuming and, with the repetition rate of at best one shot per 30 seconds, virtually impossible. In an attempt to improve on this, the electrodes of the RFQ were redesigned [11] and offline tests were carried out at MPIK in Heidelberg [12]. The modified RFQ was designed to have a bigger acceptance at the

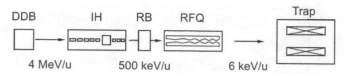

Fig. 1 Ion deceleration stages of the HITRAP facility. DDB - double drift buncher; IH - interdigital H-structure; RB - rebuncher; RFQ - radio frequency quadrupole; Trap - Penning trap

price of a larger energy dispersion of the decelerated particles. It was reinstalled at GSI and successfully commissioned in 2014. Figure 2 shows the signal of the HCI decelerated from 500 keV/u to around 6 keV/u by the RFQ, obtained after systematic scans and optimization of the system parameters. The ions leaving the RFQ were sent through the magnetic field of a permanent magnet with integrated slits. As a result, the ions with smaller energy get a larger deflection angle (the left peak in Fig. 2) and can be distinguished from the non-decelerated ions (the right peak), given with a mixture of 4 MeV/u and 500 keV/u ions. A more detailed description can be found in [13].

The future commissioning beamtimes at GSI will bring detailed analysis of the energy spectrum of the decelerated ions and their transport towards the cooling trap. Finally, the ions should be stored in the trap, where the combination of different cooling techniques brings their energy to the sub-eV range.

3 The HITRAP cooling trap

The ions leave the RFQ delecerator with a wide energy spread centred around 6 keV/u and with a very large beam emittance, making further transport very difficult. However, this energy is in principle low enough for a dynamic capture of the ions in the HITRAP cooling trap, where the ions can undergo further cooling. The trap consists of 23 gold-plated cylindrical electrodes, aligned in a row to form a 40 cm long trap. Each electrode can be supplied with high voltage individually, which can be used to create multiple regions with a quadrupole electric potential inside the electrode structure. The complete setup is situated inside the cold bore of a superconducting magnet, providing a magnetic field of up to 6 T.

The goal the cooling trap is to cool the HCI down to the temperature of the trap (4 K) or lower. Additionally, a rapid cooling mechanism is needed in order to avoid ion loss in collisions with the residual gas particles. Different cooling techniques can be used to that end; in this case a combination of electron cooling [14] and resistive cooling [15] was chosen. Evaporative cooling was not an option because it is directly connected to ion loss, possibly not fast enough and the final energy is comparably large. Sympathetic cooling with a laser-cooled ion cloud could be another option, but it was abandoned because it required laser maintenance and optical detection, which are technically demanding for an online facility with a repetition rate of one to two shots per minute. Only a limited number of experiments with resistive cooling of large clouds of HCI is available so far, indicating that the ions are slowly (of the order of a few seconds) cooled to the temperature of the electronic circuit, which can be the same or higher than the environment temperature of 4 K [16, 17]. For the current design of the cooling trap, the ion cloud will be cooled by the electrons down to some 10 eV/q, at which point the electrons should be ejected to avoid recombination, and resistive cooling should take over, cooling the ions down to the cryogenic temperatures.

 🙮 Springer

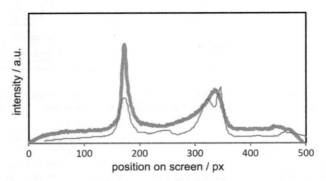

Fig. 2 Ion deceleration in the RFQ as seen by the energy analyser. The thin, red line is the reference signal from the offline tests and the thick, blue line is the online signal achieved at HITRAP. The low energy part, i.e. the decelerated ions' signal is the peak to the left. The peak to the right is the undecelerated part of the beam

First trapping tests showed the capability of electron storage, and their self-cooling through emission of synchrotron radiation. In the strong magnetic field of the trap the electrons undergo a fast cyclotron motion and experience a loss in energy via emittance of synchrotron radiation due to the high Lorentz acceleration. The time constant of this self-cooling process is of the order of a few seconds. It strongly depends on the electrons' cyclotron frequency and thereby on the magnetic field.

To prove this self-cooling behaviour, bunches of about 10^8 electrons, emitted from a GaAs surface after irradiation with UV light [18], were injected into the trap and captured between two electrodes with fast voltage switching. The electrons were ejected from the trap after different time intervals and guided to a multi channel plate (MCP) detector. The energy of the ejected electrons was measured by applying a repelling voltage to an electrode between the trap and the detector and by measuring at which field strength the electrons were fully repelled. This measurement showed an exponential decay of the electron energy as a function of the storage time. The measurement was repeated for different magnetic fields, as shown in Fig. 3, confirming the expected decrease of the electrons' synchrotron cooling time for increasing magnetic fields.

The electrode structure of the trap allows a simultaneous storage of ions and electrons in the same area. To that end, locally produced highly charged oxygen ions were injected into the trap and stored for extended amounts of time before ejecting them towards the detector. Such offline tests of the cooling trap's ion storage capability have already yielded storage times of several seconds [20]. By superimposing the ion cloud with a previously injected, self-cooled electron plasma, the ions will be able to transfer a large portion of their kinetic energy to the electrons via elastic scattering. Because of this rapid energy reduction, the storage time is expected to increase enough so that resistive cooling can take over after ejecting the electrons. Thus, the detection of electron cooling of ions is one of the main short-term objectives to be achieved with the HITRAP cooling trap. Further goals include improved vacuum conditions, the optimization of the ion and electron capture process as well as ion cooling with the resistive cooling technique down to the environment temperature of a few K.

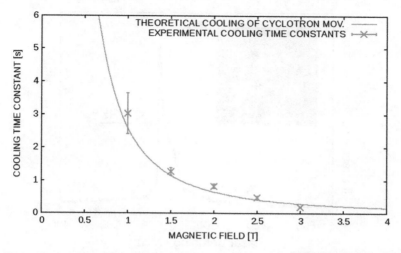

Fig. 3 The measured cooling time constants (red) in comparison with the theoretical prediction for different magnetic fields (grey). One should note that, aside from the good general agreement, the remaining deviations between experiment and theory arise from experimental parameters like field inhomogeneities and particle-particle interactions, and are not expected to vanish in the experiment [19]

4 Laser spectroscopy of highly charged ions

The HCI from the HITRAP cooling trap will be transported with an energy of around 5 keV/q towards the associated experiments. The beamline for low-energy transport of HCI was finished and commissioned in 2013. It makes a direct connection between the cooling trap, a local EBIT [21] and the HITRAP experiments, with the possibility to guide the ions in both directions. In that way, both the cooling trap (for testing purposes) and the experiments can be supplied with medium heavy HCI from an ion source independent of the GSI accelerator infrastructure. Depending on the number of ions in a bunch and their energy per charge, transport efficiencies close to 100 % were achieved [22].

As one of the experiments associated with HITRAP, the SpecTrap setup [4] is preparing to accept heavy HCI from the facility and re-trapping them in a Penning trap. With a dedicated Helmholtz-type superconducting magnet, the SpecTrap open-endcap Penning trap enables direct optical access both in the axial and radial directions. As such, it is an ideal tool for laser spectroscopy experiments with few-electron ions and the direct observation of their fluorescence.

Highly charged ions from the EBIT are transported towards SpecTrap with an energy of a few keV/q. Before trapping, this energy is reduced by a pulsed drift tube down to 500 eV/q. The ions with this energy can be trapped in the SpecTrap Penning trap, but a rapid cooling mechanism is still needed to reduce the number of charge-exchanging collisions with the residual gas particles. Cooling the ions to low temperatures also reduces the Doppler broadening of the transition frequencies, bringing a high relative accuracy of the measurement as compared to e.g. similar measurements in the GSI storage ring. Similar as in the case of the HITRAP cooling trap, both resistive cooling and sympathetic cooling are foreseen to that end. Here, laser-cooled singly charged Mg ions will be used instead of electrons. Figure 4 shows the line profile of a laser cooled Mg^+ cloud with several hundred ions.

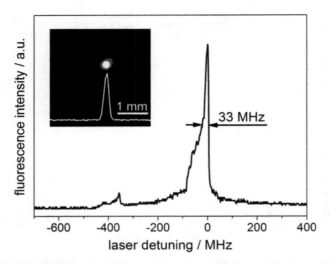

Fig. 4 Line profile of the laser cooled Mg^+ ions at SpecTrap. After the laser frequency is scanned over the transition, laser cooling turns into heating and the signal drops almost to zero. Therefore, the FWHM of the observed signal (33 MHz) represents only one half of the full profile. The comparison of the observed signal to the natural linewidth (42 MHz) can be used to give the upper limit for the ion temperature (60 mK). An image from a CCD camera is shown as inset, indicating the actual size of the cooled ion cloud

The laser wavelength was scanned across the resonance. The signal dropped rapidly to zero after crossing the central transition frequency. A comparison of the observed linewidth to the natural linewidth of the transition gives a conservative upper limit to the ion temperature of $T \leq 60$ mK [4]. The inset shows the cooled ion cloud recorded with a CCD camera, indicating its size of less than a millimeter.

Using the ions produced locally by the EBIT, the laser cooled ions in SpecTrap were mixed with HCI and the investigation of the ion dynamics in ongoing. Due to the low temperature achievable through sympathetic cooling, the expected relative accuracy of laser spectroscopy of forbidden transitions in HCI is of the order of $10^{-7} - 10^{-8}$. The limiting factors are the ion lifetime in the trap and the availability of the suitable laser systems and fluorescence detectors. Taking that into account, the measurement candidates include fine structure transitions in medium-heavy systems like Ar^{13+} and Ca^{14+} produced locally, as well as hyperfine structure transitions in $Bi^{80+,82+}$ produced by the GSI accelerator complex and decelerated by the HITRAP facility [23]. Alternatively, highly charged heavy ions can be extracted from the so-called S-EBIT [24], on loan from Helmholtz Institute Jena and currently under construction at HITRAP. It will be connected directly to the existing HITRAP infrastructure and together with the existing EBIT, it will provide medium heavy and heavy highly charged ions, independent of the accelerator beamtime at GSI, thus helping to bridge the shutdown period necessary for the construction of FAIR [25].

5 Conclusions

As the first facility of its kind in the world, HITRAP has had to overcome many difficulties on the road towards large clouds of heavy highly charged ions decelerated all the way from the their production energy of 400 MeV/u down to the sub-eV range. The first promising results have shown that the desired multi-stage deceleration can work as a concept, but requires fine tuning specific to the application. The ions were nevertheless successfully decelerated down to about 6 keV/u which makes dynamic capture in a Penning trap possible. The combination of high vacuum, high voltage, strong magnetic field and cryogenic environment makes also this step very demanding and it will require more work in the coming years. However, the trap itself has been tested with locally produced ions and electrons, including the demonstration of the synchrotron cooling necessary for sympathetic electron cooling of HCI.

The experiments around HITRAP are developed in parallel to the facility and have also seen first tests with locally produced ions. One of them is SpecTrap, where singly charged Mg ions were loaded into the trap and laser cooled to a temperature significantly under 1 K. As such, they will enable sympathetic cooling of HCI which may come from a local EBIT or the HITRAP facility. The first test with Ar^{13+} have already been carried out and further optimization of the process is ongoing. The long term goal of the experiment is trapping and laser spectroscopy with $Bi^{80+,82+}$. The energies of the hyperfine splitting in these two ion species have been recently measured in the ESR [8] and an increase in relative accuracy of several orders of magnitude is expected in the trap.

References

1. Gumberidze, A., et al.: Quantum electrodynamics in strong electric fields: The ground-state Lamb shift in hydrogenlike uranium. Phys. Rev. Lett. **94**, 223001 (2005)
2. Shabaev, V.M., Artemyev, A.N., Yerokhin, V.A., Zherebtsov, O.M., Soff, G.: Towards a test of QED in investigations of the hyperfine splitting in heavy ions. Phys. Rev. Lett. **86**(18), 3959 (2001)
3. Kluge, H.-J., et al.: Advances in Quantum Chemistry **53**, 83 (2008)
4. Andelkovic, Z., et al.: Laser cooling of externally produced Mg ions in a Penning trap for sympathetic cooling of highly charged ions. Phys. Rev. A **87**, 033423 (2013)
5. von Lindenfels, D., et al.: Experimental access to higher-order Zeeman effects by precision spectroscopy of highly charged ions in a Penning trap. Phys. Rev. A **87**, 023412 (2013)
6. Götz, S., Höltkemeier, B., Hofmann, C.S., Litsch, D., DePaola, B.D., Weidemüller, M.: Versatile cold atom target apparatus. Rev. Sci. Instrum. **83**, 073112 (2012)
7. Vogel, M., Quint, W., Paulus, G.G., Stöhlker, Th.: A Penning trap for advanced studies with particles in extreme laser fields. Nucl. Instr. Meth. Phys. Res. B **285**, 65 (2012)
8. Lochmann, M., et al.: Observation of the hyperfine transition in lithium-like bismuth $^{209}Bi^{80+}$: Towards a test of QED in strong magnetic fields. Phys. Rev. A **90**, 030501(R) (2014)
9. Beiersdorfer, P., Chen, H., Thorn, D.B., Träbert, E.: Measurement of the two-loop Lamb shift in lithiumlike U^{89+}. Phys. Rev. Lett. **95**, 233003 (2005)
10. Herfurth, F., et al.: HITRAP - Heavy, Highly-Charged Ions at Rest - A Status Report, GSI Scientific Reports (2011)
11. Yaramyshev, S., et al.: A new Design of the RFQ-Decelerator for HITRAP, GSI Scientific Reports (2012)
12. Maier, M., et al.: Offline commissioning of the old and new HITRAP RFQ, GSI Scientific Reports (2012)
13. Herfurth, F., et al.: The HITRAP facility for slow highly charged ions, Proceedings of the STORI 2014 conference, accepted for publication in Phys. Scr. (2014)
14. Zwicknagel, G.: Electron cooling of ions and antiprotons in traps. AIP Conf. Proc. **821**, 513 (2006)
15. Wineland, D.J., Dehmelt, H.G.: Principles of the stored ion calorimeter. J. Appl. Phys. **46**, 919 (1975)
16. Häffner, H., et al.: Double Penning trap technique for precise g factor determinations in highly charged ions. Eur. Phys. J. D **22**, 163 (2003)
17. Gruber, L., Holder, J.P., Schneider, D.: Formation of strongly coupled plasmas from multi-component ions in a Penning trap. Phys. Scr. **71**, 60 (2005)
18. Krantz, C.: Intense electron beams from GaAs photocathodes as a tool for molecular and atomic physics, PhD Thesis, University Heidelberg (2009)

19. Maaß, B.: Geladene Teilchen in der Kühlfalle, Master's thesis, TU Darmstadt (2014)
20. Fedotova, S.: Experimental characterization of the HITRAP cooler trap with highly charged ions, PhD Thesis, University Heidelberg (2013)
21. Sokolov, A., Herfurth, F., Kester, O., Stoehlker, Th., Thorn, A., Vorobjev, G., Zschornack, G.: SPARC EBIT - a charge breeder for the HITRAP project, JINST 5 C11001 (2010)
22. Andelkovic, Z., et al.: Beamline for low-energy transport of highly charged ions at HITRAP. Nucl. Instr. Meth. Phys. Res. A **795**, 109 (2015)
23. Schmidt, S., Geppert, Ch., Andeklovic, Z.: Laser spectroscopy methods for probing highly charged ions at GSI. Hyp. Int. **227**, 29 (2014)
24. Schuch, R., et al.: The new Stockholm Electron Beam Ion Trap (S-EBIT). J. Instrum. **5**, C12018 (2010)
25. http://www.fair-center.de/ (2015)

Hyperfine Interact (2015) 235:45–49
DOI 10.1007/s10751-015-1195-z

Intensity ratio measurements for density sensitive lines of highly charged Fe ions

Safdar Ali[1] · Erina Shimizu[1] · Hiroyuki A. Sakaue[2] · Daiji Kato[2,3] ·
Izumi Murakami[2,3] · Norimasa Yamamoto[4] · Hirohisa Hara[3,5] ·
Tetsuya Watanabe[3,5] · Nobuyuki Nakamura[1]

Published online: 8 July 2015
© Springer International Publishing Switzerland 2015

Abstract Intensity ratio of density sensitive emission lines emitted from Fe ions in the extreme ultraviolet region is important for astrophysics applications. We report high-resolution intensity ratio measurements for Fe ions performed at Tokyo EBIT laboratory by employing a flat-field grazing incidence spectrometer. The experimental intensity ratios of Fe X and Fe XII are plotted as a function of electron density for different electron beam currents. The experimental results are compared with the predicted intensity ratios from the model calculations.

Keywords Atomic data · Spectroscopy · Plasmas · Electron density

1 Introduction

Spectroscopic investigations of density sensitive lines emitted from highly charged ions play an important role to measure the properties of astrophysical and laboratory

Proceedings of the 6th International Conference on Trapped Charged Particles and Fundamental Physics (TCP 2014), Takamatsu, Japan, 1-5 December 2014

✉ Nobuyuki Nakamura
n_nakamu@ils.uec.ac.jp

Safdar Ali
safdaruetian@gmail.com

[1] Institute for Laser Science, The University of Electro-Communications, Tokyo 182 8585, Japan

[2] National Institute for Fusion Science, Gifu 509-5292, Japan

[3] The Graduate University of Advanced Studies (SOKENDAI), Gifu 509-5292, Japan

[4] Chubu University, Aichi 487-8501, Japan

[5] National Astronomical Observatory of Japan, Tokyo, 181-8588, Japan

plasmas. In particular, emission line ratios in the extreme ultraviolet (EUV) spectral region provides an essential tool to study the most violent phenomena-taking place in different astrophysical objects. To understand the structure of solar corona, for example, one needs to know temperatures, densities, and elemental abundances, which can be obtained by having the knowledge of spectroscopy. In the coronal temperature range (1−3 MK), emission features originating from transitions in Fe X, Fe XI, Fe XII ions are considered to be the main source of density sensitive lines [1]. These originate mainly below 300 Å and many of them are useful for density determination of the astrophysical plasma. This wavelength region is very well covered with the EUV Imaging Spectrometer (EIS) on board the Hinode satellite [2] and provides motivation to perform laboratory measurements for the modeling and interpretation of the observed data.

The sensitivity on electron density in the range of high temperature solar and flare plasmas can be tested in a laboratory with several methods. The most direct and convenient method is based on line intensity ratio measurements [3] and electron beam ion traps (EBITs) are very well suited for these type of studies. To extend our ongoing efforts to derive Fe spectroscopic data [4], in this paper, we present high-resolution density sensitive intensity ratio data of highly charged Fe X and Fe XII ions in EUV wavelength range obtained with a compact low-energy EBIT [5, 6]. The derived results will be presented and compared with the model calculations.

2 Experiment

The EUV emission spectroscopy measurements reported here were performed at a compact electron beam ion trap called CoBIT developed at the University of Electro-communications [2] Tokyo, Japan. To produce desired Fe charge states, $Fe(C_5H_5)_2$ vapor was injected into CoBIT through a variable leak valve gas injection system, while keeping the vacuum in the EBIT chamber below 10^{-8} Pa. The ions produced as a result of electron collisions with the injected vapor were trapped by applying a potential of 30 V on the outer drift tubes. The EUV emission spectra were recorded with the aid of a high-resolution grazing incidence flat-field grating spectrometer [7]. In contrary to our previous studies [4] a concave grating having a larger radius of curvature (Hitachi 001-0660) to obtain higher dispersion is used in the present measurements. Although the average groove number is the same as that of the previous flat field grazing incidence spectrometer grating [7], but the larger radius of curvature (13450 mm) and the larger distance from the grating to the focal plane (563.2 mm) makes the dispersion on the focal plane higher as 2.6 Å/mm. In the present setup, the spectrometer is used in the still less configuration (no entrance slit) because an EBIT represents a thin line shape source. The spectral resolution in the present measurements was typically 0.4 Å, which is very much improved than our previous reported measurements, where it was 0.8 Å.

The spectroscopic data was recorded with an automatic data acquisition system and stored counts and channel numbers in the computer. Figure 1 displays typical data (background subtracted) plots as a function of wavelength obtained at an electron beam energy of 340 eV with different values of electron beam currents. Each data set was recorded for exposure time of 30 min. The dominant charge states are Fe X, Fe XI, and Fe XII as evident from the Fig. 1. The trap was emptied periodically to avoid accumulation of heavy impurity ions such as W and Ba evaporated from the cathode. Indeed, the obtained spectra contain no sign of any line from heavy impurities in our measurements. The pixel to wavelength scale was calibrated using six well-known Fe lines [Fe IX (171.073 Å), Fe XI (180.401, 188.216

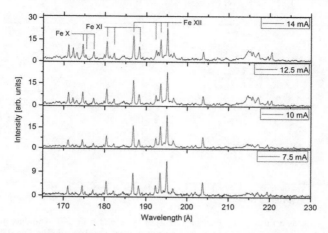

Fig. 1 EUV spectra of highly charged Fe ions obtained with CoBIT at electron beam energy of 340 eV with different electron beam currents

Å), Fe XIII (203.826 Å), Fe XIV (211.317, 219.130 Å)] by fitting the calibration curve with a third degree polynomial.

3 Results and discussion

The EUV spectra collected from highly charged iron ions is shown in Fig. 1 in the wavelength range of 160 to 230 Å. In this wavelength range, the prominent lines are found to stems from Fe X, Fe XI and Fe XII, while at high wavelength the most dominant lines are seen to be from high charge states. We also recorded spectra for higher energies such as 400 and 500 eV and found that the intensity of low charge states decreases while increase for higher charge states such as Fe XIII and Fe XIV. This shows that the plasma inside the trap has a narrow charge state distribution, which is very important to obtain spectra free from any kind of overlap or blends.

As discussed earlier emission feature arising from Fe X, Fe XI and Fe XII transitions are important to determine density of the hot plasma existed in astrophysical environment. Almost all the lines identified in our measured spectra of highly charged iron have the density diagnostics potential; in this paper, we will restrict our discussion about the line ratios corresponding to Fe X transitions $3s^2 3p^5\ {}^2P_{3/2}$-$3s^2 3p^4 ({}^3P) 3d\ {}^2P_{3/2}$ at 177.243 Å and $3s^2 3p^5\ {}^2P_{3/2}$-$3s^2 3p^4 ({}^3P) 3d\ {}^2D_{3/2}$ at 175.263 Å and Fe XII transitions $3s^2 3p^3\ {}^2D_{5/2}$-$3s^2 3p^2 ({}^3P) 3d\ {}^2F_{7/2}$ at 186.887 Å and $3s^2 3p^3\ {}^4S_{3/2}$-$3s^2 3p^2 ({}^3P) 3d\ {}^2D_{5/2}$ at 195.119 Å. A detail discussion about all other density sensitive lines ratio corresponding to Fe X, Fe XI and Fe XII (labeled in Fig. 1) will be reported in our upcoming publication [8].

The Fe X line at 174.534 Å is considered to be the strongest line in the spectra of Procyon and Alpha Centauri A and does not suffer from any blends from other lines. The second line of present interest at 175.263 Å might blend with Fe X 175.474 Å ($3s^2 3p^5\ {}^2P_{3/2}$-$3s^2 3p^4 ({}^3P) 3d\ {}^2P_{1/2}$) but its contribution is negligible (about 10 %) [9, 10]. Several Fe XI and Fe XII lines are observed with EIS spectrometer [11]. The strongest line from Fe XII at 195.119 Å lies at the peak of the EIS sensitivity curve and is well identified. There is another Fe X II transition at 195.180 Å, which may contribute to the emission line 195.119 Å. Below electron beam density of 10^{10} cm^{-3} the contribution of this line is predicted to be

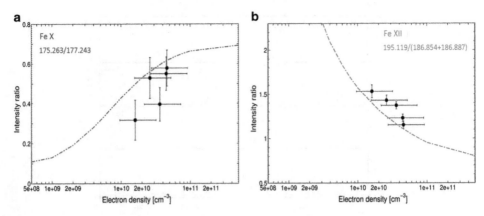

Fig. 2 Intensity ratio as a function of electron density **a** Fe X and **b** Fe XII. The data points with error bar are the experimental values obtained for electron beam energy of 340 eV. The five data point in each figure corresponds to electron beam current of 7.5, 10, 12.5 and 14 mA starting from *lower* to high density, respectively. The *dotted dash-lines* correspond to the model calculations

≤ 10 % by CHIANTI, while at higher density 10^{11} cm^{-3} this contribution enhances up to 22 % [11]. The second line of interest from Fe XII at 186.887 Å is blended with another Fe XII line 186.854 Å. The Fe X II line ratios, particularly 195.119 Å/186.887 Å is very sensitive to the electron density in the solar corona and gives the best density diagnostic due to its broad range of sensitivity [12].

The density dependent line emissivity is simulated by using collisional-radiative (CR) models. The CR models give fractional population of ions in excited states, n_i, at given electron energies and densities, n_e, by solving quasi-stationary-state rate equations for the fractional population,

$$0 = \sum_j \left(A_{ij} + n_e C_{ij}\right) n_j - \left(\sum_j A_{ji} + n_e C_{ji}\right) n_i - n_e S_i n_i,$$

where A_{ij} stands for a spontaneous emission rate from the upper level j to the lower level i, C_{ij} electron collision rate coefficients between the levels, and S_j ionization rate coefficients from the level i. In the present model, the spontaneous transitions of electric-dipole, -quadruple, and -octupole, and those of magnetic-dipole and -quadruple are taken into account. The collision rate coefficients and the ionization rate coefficients are calculated from electron-impact excitation and ionization cross sections, respectively, assuming the delta function of electron energy for the electron energy distribution in the CoBIT.

We constructed the CR model for Fe X and Fe XII using the atomic data obtained from HULLAC code [13]. In the present model, electronic configurations of $3s^2 3p^4 (3d, nl)$, $3s 3p^6$, and $3s 3p^5 (3d, nl)$ for the excited states of Fe X, and $3s^2 3p^2 (3d, nl)$, $3s 3p^4$, and $3s 3p^3 (3d, nl)$ for those of Fe XII are included, respectively, where $n = 4, 5$ and $l \leq n - 1$. Additional electronic configurations that differ by two electrons from $3s^2 3p^{k-1} 3d$ but the same parity, where $k = 3$ and 5 for Fe XII and Fe X, respectively, are also included. This augmentation significantly improved the wavelengths of the present interest lines. For Fe X, the ground state, $3s^2 3p^5$, is also augmented by including $3s^2 3p^3 3d^2$.

The experimental intensity ratio for the line pairs mentioned above is shown in Fig. 2 as a function of electron beam density. An electron beam imaging set up was used to obtain the spatial distribution of the EUV emission. This emission is considered to represent the electron density distribution since the lifetime of EUV transitions is of the order of $\sim 10^{-10}$s.

More detail of the electron beam density determination can be found in [4]. The experimental data points obtained with an electron beam energy of 340 eV, with different electron beam currents are shown by closed circles.

As can be seen from Fig. 2a and b, qualitative agreement between the present experiment and theory is found, but the experimental points seem to have a tendency to shift to the higher density side. One obvious reason for this disagreement might be the wrong estimation of the overlap between the electron beam and the ion cloud. In the present data analysis, we have used the size of ion cloud equal to electron beam, which might not be true in reality since the ion cloud size is usually larger than the electron beam size. Another reason for this discrepancy could be the strong dependence of ion cloud size on the charge states as discussed by Liang et al. [14]. A complete description of the charge state dependence on electron density and intensity ratios as a function of electron density for different Fe ions will be given in our upcoming publication [8].

4 Conclusions

We have reported electron density sensitive intensity ratio results for Fe X and Fe XII ions important for astrophysical plasma applications. The experimental results are compared with the calculations and qualitative agreement is found although a tendency that the experimentally determined density is higher than the theoretical density. We consider that the disagreement between the present calculated and experimental results might be due to the underestimation of ion cloud size, which resulted in high electron beam density reported here. Further study on electron densities and line ratios for these ions is currently underway and will be reported very soon in our upcoming publication.

Acknowledgments   This work was performed under the Research Cooperation Program in the National Institutes of Natural Sciences (NINS).

Conflict of interests   The author declares that there is no conflict of interests regarding the publication of this paper.

References

1. Young, P.R., Watanabe, T., Hara, H., Mariska, J.T.: A&A **495**, 587 (2008)
2. Culhane, J., Harra, L.K., James, A.M., et al.: Sol. Phys. **243**, 19 (2007)
3. Feldman, U., Landi, E., Doschek, G.A.: APJ **679**, 843 (2008)
4. Nakamura, N., Watanabe, E., Sakaue, H.A., Kato, D., Murakami, I.: APJ **739**, 17 (2011)
5. Sakaue, H.A., Nakamura, N., Watanabe, E., Komatsu, A., Watanabe, T.: J. Instrum. **5**, C08010 (2010)
6. Nakamura, N., Kikuchi, H., Sakaue, H.A., Watanabe, T.: Rev. Sci. Instrum. **79**, 063104 (2008)
7. Sakaue, H.A., Kato, D., Nakamura, N., et al.: J. Phys.: Conf. Ser. **163**, 012020 (2009)
8. Ali, S., et al: to be submitted to APJ (2015)
9. Brickhouse, N.S., Raymond, J.C., Smith, B.W.: APJS **97**, 551 (1995)
10. Foster, B.J., Mathioudakis, M., Keeenan, F.P., et al.: APJ **473**, 560 (1996)
11. Young, P.R., Del Zanna, G., Mason, H.E., et al.: PASJ **59**, 857 (2007)
12. Pradeep, K., Bhupendra, S., Anil, K.: Res. J. Recent Sci. **1**, 40 (2012)
13. Bar-Shalom, A., Klapisch, M., Oreg, J.: J. Quant. Spectrosc. Radiat. Transfer **71**, 169 (2001)
14. Liang, G.Y., Crespo López-Urrutia, J.R., Baumann, T.M.: APJ **702**, 838 (2009)

◯ Springer

Hyperfine Interact (2015) 235:61–75
DOI 10.1007/s10751-015-1191-3

High-resolving mass spectrographs and spectrometers

Hermann Wollnik[1]

Published online: 22 May 2015

Abstract Discussed are different types of high resolving mass spectrographs and spectro-meters. In detail outlined are (1) magnetic and electric sector field mass spectrographs, which are the oldest systems, (2) Penning Trap mass spectrographs and spectrometers, which have achieved very high mass-resolving powers, but are technically demanding (3) time-of-flight mass spectrographs using high energy ions passing through accelerator rings, which have also achieved very high mass-resolving powers and are equally technically demanding, (4) linear time-of-flight mass spectrographs, which have become the most ver-satile mass analyzers for low energy ions, while the even higher performing multi-pass systems have only started to be used, (5) orbitraps, which also have achieved remarkably high mass-resolving powers for low energy ions.

Keywords Accurate mass determinations of ions · Masses of short-lived nuclei ·
Molecule identification through their masses

1 Introduction

The goal of mass spectroscopy is to distinguish ions whose masses m_0 and $m = m_0 (1 + \delta_m)$ differ only by small percentages δ_m. There are mass spectrographs [1, 2] in which ions of different masses are recorded simultaneously in a position sensitive ion detector, but there are also mass spectrometers [3] in which the intensities of ions of different masses are recorded behind the same exit slit, when the electromagnetic fields in the mass analyzer are different. Such a mass spectrometer is an easier to operate system but features lower transmissions T_m since at one instant only ions of one specific mass are recorded.

Proceedings of the 6th International Conference on Trapped Charged Particles and Fundamental
Physics (TCP 2014), Takamatsu, Japan, 1-5 December 2014

✉ Hermann Wollnik
  hwollnik@gmail.com

[1] Department of Chemistry & Biochemistry, New Mexico State University, Las Cruces, NM, USA

 Springer

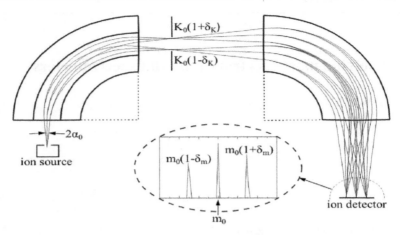

Fig. 1 An angle- and energy-focusing mass spectrograph is sketched that separates equally charged ions of energies $K = K_0(1 \pm \delta_K)$ and masses m_0 and $m_0(1 \pm \delta_m)$ according to their masses only. independent of the ions relative energy deviations δ_K and the initial inclinations $\alpha \leq \alpha_0$ of the ion trajectories. Please note also that, if ion source and ion detector would be exchanged, the mass spectrograph would still be angle- and energy-focusing [4] and the mass resolving power would be the same. However, the mass dispersion would be different and so would the lateral magnification

The performance of mass analyzers is best characterized by the product of their ion transmission T_m and of their mass-resolving power R_m, which is the inverse of the smallest resolvable $(\delta_m)_{min}$. Problematic is always that ions are not produced with one specific kinetic energy K_0, but with a range of energies $K = K_0 (1 + \delta_K)$. This energy spread must be compensated in order that a pure mass dispersion is achieved [1, 2, 4].

There are two main applications of mass spectrographs and mass spectrometers that require very high mass-resolving powers R_m.

- The determination of masses of atoms, which provides direct information on nuclear binding energies and for the understanding of their nuclear structure [5, 6]. Together with measured nuclear half-lives [6, 7] precise mass measurements are also important to understand the nucleosynthesis in stars [8].
- The determination of masses of molecules, which allows to identify and to distinguish complex molecules and to determine their composition in medical, pharmacological, biological and/or environmental investigations [9, 10].

2 Angle- and energy-focusing sector-field mass-spectrographs

Ions of different charges q, masses $m = m_0 (1 + \delta_m)$, and energies $K = K_0 (1 + \delta_k)$ are deflected [4] in magnetic sector fields of strength B_0 along radii $\rho_B = \sqrt{2mK}/q B_0$ and in electric sector fields of strength E_0 along radii $\rho_E = K/q E_0$. Combining at least one magnetic and one electric sector field, it is possible to achieve [2, 4] that the ion arrival positions at a final ion detector depend only on the mass/charge ratio m/q of the investigated ions and that these are independent of the ions' energy spreads $K_0\delta_K$ as well as of their initial angles of inclination $\alpha \leq \alpha_0$ of their trajectories in the plane of deflection. For $\delta_K \ll 1$ and $\alpha_0 \ll 1$ such systems feature a pure mass dispersion along an image plane as is illustrated in Fig. 1.

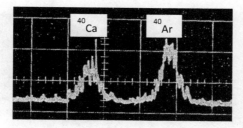

Fig. 2 A highly resolved mass doublet, whose relative mass difference is only 207 μu or ≈5.2 ppm of mass 40, obtained by a sector field mass spectrograph [12]

Such laterally angle- and energy-focusing mass analyzers commonly achieved FWHM mass-resolving powers of up to a few 1,000 during the 1920s. Higher mass-resolving powers were reached, when the mechanical precision of the systems was improved allowing FWHM mass-resolving powers of $R_m \approx 26,000$ [11] already in the 1940s and in the 1960s FWHM mass-resolving powers [12] of $R_m \approx 1,000,000$ (see Fig. 2). However, the ion transmissions T_m of these systems were rather limited, i.e. $\alpha \ll 1$ and $\delta_K \ll 1$ were mandatory. More fundamental were the severe limitations due to image aberrations, which changed the arrival positions of the ions at the detector. The most important limitations were here the second-order aberrations, the largest of which usually were the terms proportional to α_0^2, δ_K^2, $\alpha_0 \delta_K$. Mass analyzers that achieve simultaneously high mass-resolving powers and high ion transmissions could only be built, when the image aberrations were reduced. This required to understand not only the ion motion within the magnetic and electric sector fields [2] but also their complex motion through the fringing fields of these sector fields [4, 13].

3 Penning trap mass spectrographs and spectrometers

A very different way to determine the masses of ions is to record the frequencies ω of their circular motion in a homogeneous magnetic field B_0[5]. This so-called cyclotron frequency $\omega = q B_0/m$ depends on the ions' mass/charge ratios m/q, but is independent of the initial angles of inclination $\alpha \le \alpha_0$ of the ions' trajectories and most importantly of the ions' energies K since both the velocities $v = \sqrt{2K/m}$ as well as the radii of deflection $\rho_B = \sqrt{2mK}/q B_0$ of the ions increase with the square root of the ion energies and since $\omega = 2\pi \rho_B/v$. Having recorded this cyclotron frequency ω it must be compared to a precision reference frequency, wherein obviously the accuracy of this comparison improves, if the experiment is stretched out to as many ion rotation cycles as possible. If this measurement is to be performed in a limited time, the frequency ω should be chosen as high as possible, which requires to choose the magnetic field B_0 as well as the ion charge q to be as high as possible. In all cases it is found that the achievable mass accuracy is proportional to $1/m$ showing that the accuracy of a mass measurement and so the mass-resolving power of a Penning trap mass analyzer decreases linearly with increased ion masses.

Though the ion motion in the plane of deflection is confined in a magnetic field there is no confinement in the perpendicular direction, the so-called axial direction. Such an axial confinement can be achieved, however, by establishing a superimposed electrostatic field (see Figs. 3 and 4) that drives the ions back to the mid-plane of the magnet, should they ever deviate from this plane. In order to start the motion of ions in this mid-plane, however, the

Springer

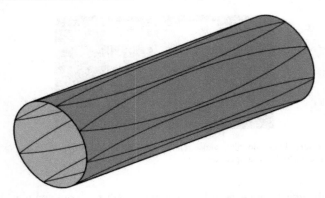

Fig. 3 Illustrated is one of the possible electrode arrangements that forms the desired back-driving electric field in a non-destructive Penning Trap mass or FT-ICR spectrograph to a mid-plane. This cylinder is submerged in the magnetic field so that the axis of the electrodes is parallel to the magnetic field lines and if different DC-potentials are applied to the spindle-like and to the triangular-like electrodes, an electric field is formed that increases linearly with the distance from the mid-plane of the cylinder. In this mid-plane, thus, ions can cycle in the magnetic field perpendicularly to the axis of the cylinder and are driven back to this mid-plane whenever ions should deviate from it

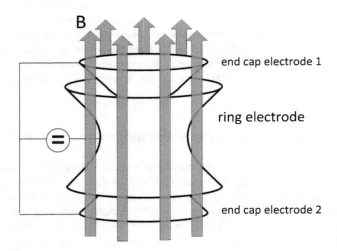

Fig. 4 Illustrated are three electrodes that form an electrostatic rotationally symmetric DC quadrupole field submersed in the homogeneous magnetic field of a Penning trap with the axes of the three electrodes coinciding with the magnetic field lines. Due to the electric field in this quadrupole the ions are driven back to the ring electrode's mid-plane

ions must be injected into the Penning trap with almost vanishing energy, which requires special efforts.

If at least one of these electrodes is sectioned azimuthally, one can record signals induced by the cycling ions in some of these sections, which, after a Fourier transform, directly reveal the ion cycling frequencies and thus the masses of the ions. However, it is also possible to establish a high frequency excitation field between neighboring electrode sections, which, depending on the excitation frequency, will cause only ions of a corresponding mass/charge ratio to gain energy and so increase the radii of their trajectories.

3.1 Ion non-destructive penning trap mass spectrographs

In one version of a Penning trap mass analyzer, ions of many masses cycle simultaneously in the systems [14] with different frequencies ω. In such systems the axial confinement field is formed [15] by charged "printed circuit electrodes" placed for instance on the surface of a cylinder (see Fig. 3) submerged in the magnetic field, wherein the cylinder axis coincides with the axis of the magnetic field. To these spindle-like and triangle-like "printed circuit electrodes" different DC-potentials can be applied that form an electric field, which increases linearly with the distance from the magnet mid-plane and thus drives the ions back to this plane. When the ions cycle in the magnetic field B_0 within this cylinder, they will induce high frequency signals in neighboring electrodes, which after a Fourier analysis will reveal a spectrum of cyclotron frequencies and thus the mass spectrum of the cycling ions.

Such Penning trap mass analyzers are often also referred to as "Fourier Transform Ion Cyclotron Resonance" or FT-ICR mass spectrographs. The mass-resolving power of such systems is high. Even for ions of masses of $\approx 1,000$ u mass-resolving powers of about one million are achieved [16]. This enables such FT-ICR mass spectrographs to identify molecules also in complex mixtures that contain ensembles of ions of many neighboring masses [17]. In such FT-ICR mass spectrographs Coulomb forces between the circling ions may cause frequency shifts and thus errors in the determination of absolute ion masses. However, as long as these effects stay within limits, the achievable mass resolving power is substantially unchanged [18].

3.2 Ion-destructive penning trap mass spectrometers

Another version of a Penning trap mass analyzer has been developed to determine the masses of single or very few ions of the same mass in which case the Coulomb forces between the circling ions are small or nonexisting so that the masses of ions can be determined with ultimate precision [5]. In such a system the axially confining electrostatic DC-field is commonly achieved by rotationally symmetric quadrupole electrodes shaped as one ring electrode and two end cap electrodes (see Fig. 4). This electrode arrangement is submerged in the magnetic field B_0 in such a way that the axis of symmetry of the electrodes is parallel to the magnetic field. The ring electrode in such a system is usually divided azimuthally in different sections between at least some of which additional high frequency potentials can be applied whose fields excite a cycling ion to higher energies so that the ion moves along an increased radius.

The exact value of the resonant frequency ω here is determined as that one for which the ion under consideration has absorbed the highest energy during the excitation process. To determine this ω the ion under consideration is excited in a number of consecutive measurements, wherein in the different measurements slightly different frequencies are applied to the azimuthal sections of the ring electrode, with the goal to determine for which frequency the ion takes up the highest azimuthal energy. The magnitude of these excitations is found from accelerating the cycling ions in the direction of the magnetic field and through its fringing field, which causes the ion's azimuthal energy is converted into axial energy that can be determined from the ion's flight time to an external detector. In this so-called TOF-ICR technique [19, 20] the excitation frequency ω is applied for a certain time defined by a certain number of cycles of a reference frequency and by determining the number of cycles of ω during this period. To improve the accuracy of this technique   it is advantageous to determine the actual position of the ion under consideration along its circular trajectory from where the ion has been extracted and at which position the ion was, when the excitation frequency had been started [21] by using a position sensitive time-of-flight detector for the extracted ions.

By using the TOF-ICR technique mass precisions of a few 10^{-9} have been achieved [5], which for an ion of mass $\approx 100u$ corresponds to $\approx 0.1\mu u$. Such precisions are achieved, when only ≤ 100 ions have been investigated. However, the life times of these ions must be longer or at least comparable to the overall ion cycling time, which usually requires life times of the investigated ions of $\geq 100ms$. Usually such nuclei are produced in high energy nuclear reactions but must be injected into a Penning Trap at very low energies. Thus they must be slowed down in solid and/or gaseous energy absorbers and then a retarding electric field or only in a retarding electric field in case they were formed in a solid from where they diffused out, were ionized and mass separated in a magnetic isotope separator at keV energies.

In the precise mass measurement of short-lived nuclei it is usually difficult to provide reference ions of exactly known masses. This is especially so, if superheavy nuclei are to be investigated, as their masses are heavier than any stable nuclei. As a way out one can [5] use ^{12}C-clusters as reference ions or molecular ions [22] that can be produced for all isobars.

4 Time-of-flight mass analyzers

Ions of masses $m_0(1 + \delta_m)$ and energy $K_0 (1 + \delta_K)$ move through a flight path of length L with a velocity $v = \sqrt{2K_0(1 + \delta_K)/m(1 + \delta_m)}$ in a time $T(1 + \delta_T) = L/v$ where $\delta_T \approx \delta_m/2$. Thus the mass-resolving power and the energy resolving power $R_K = 1/(\delta_K)_{min}$ are both about half as large as the time resolving power $R_T = 1/(\delta_T)_{min}$. Mono-mass ions of energies $K_0(1 + \delta_K)$ pass through such systems in times $T_0[1 + (T|\delta)\,\delta_K + (T|\delta\delta)\,\delta_K^2 + (T|\delta\delta\delta)\,\delta_K^3 + \cdots]$ including aberrations to third order in δ_K[4, 23, 24]. Time-of-flight systems for which at least $(T|\delta_K) = 0$ are called energy-isochronous, which is achieved by using electromagnetic fields that send ions of higher energies along longer flight paths [25–27].

- For the determination of masses of high-energy ions start- and stop-detectors can be placed at the beginning and at the end of such energy-isochronous time-of-flight mass analyzers. In most cases such timing detectors consist of thin foils through which the ions must pass releasing on both foil surfaces secondary electrons, which can be recorded as timing signals [28].
- For the determination of masses of low-energy ions such detectors can only be used as stop detectors, while the start time, is best determined from the time, when pulses of ions are extracted from ion-storage devices [29].

4.1 Time-of-flight mass spectrographs for high energy ions

One way to build an energy isochronous time-of-flight mass analyzer is to use magnetic and/or electric sector fields. Such systems can be designed so [29] that ions of different mass/charge (m/q) and energy/charge (K/q) ratios as well as different trajectory inclinations α in the plane of deflection and β perpendicular to it are spatially focused to the same position, while the ion flight times depend only on m/q but not on K/q and not on α or β. Such a system has been built [30, 31] consisting of four identical and symmetric sector

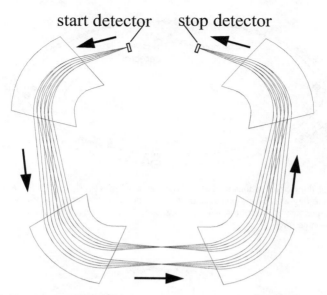

Fig. 5 Shown is the first time-of-flight mass spectrograph for energetic heavy ions in which to second order all ions are focused to a point and their flight times depend only on the ions' mass/charge ratio. Note that trajectories of ions of two energies are shown, wherein the ions of higher energies are moving on flight paths that are so much longer than the flight paths of ions of lower energies that the ion flight times are identical

magnets as sketched in Fig. 5, which provided good first and second order position and time focusing.

Obviously the flight time and the mass-resolving powers increase with the length of the flight distance in an energy isochronous time-of-flight mass spectrograph. Thus, it is advantageous to build large systems or to use a given flight path repeatedly. This is the case, when the ion flight paths are the ion trajectories in an accelerator storage ring and when the ion optics of this ring is tuned to be energy isochronous [32], in which case the ion flight time per lap depends only on the ions' mass/charge ratios but not on their energies. For the accelerator storage ring ESR at the GSI in Germany this condition could be achieved by simply changing the excitations of the quadrupoles in the ring [32].

In such an energy isochronous storage ring an actual ion's mass/charge measurement can be performed by determining the difference between the time T_1, when an ion is injected into such a ring, and the time T_2, when after a certain number of laps the ion is recorded in some detector after this ion has been ejected from the ring [33]. Another possibility is to locate in the energy isochronous storage ring a thin foil (see Fig. 6) and record the released secondary electrons [34, 35] after every lap of the ion under consideration. Such timing signals can be recorded for a few hundred and in some cases a few thousand laps [36, 37] of a cycling ion, before it is scattered out of the phase space of the ion beam. As long as these ions survive a sufficient number of laps their mass/charge ratios can be determined [6] with FWHM mass-resolving powers of up to $R_m \approx 200,000$.

When the ion-optical parameters of a storage ring are adjusted to be energy isochronous for ions of mass m_0 the ions' flight time per lap depends only on the ions' mass m_a and their mean energy K_a but not on small deviations $\pm \Delta K_a$ of individual ions from the mean energy K_a. For the same ion-optical parameters of the storage ring the flight time per lap of

© Springer

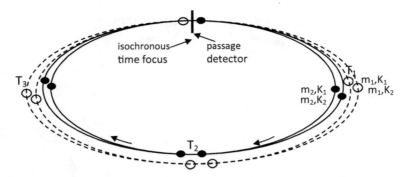

Fig. 6 Sketch of the motion of ions of masses $m_1 m_2$ and energies $K_1 K_2 = K_1 + \Delta K$ for one lap in an accelerator storage ring whose ion-optical parameters are chosen to be energy isochronous. In such a system the flight time separation of ions of masses $m_1 m_2$ increases by the same amount for every lap at any chosen position in the storage ring

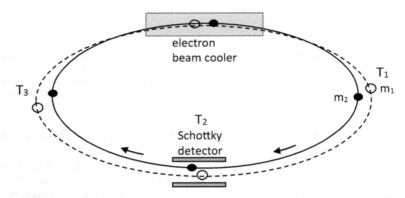

Fig. 7 Sketch of the motion of ions of masses $m_1 m_2$, which have been cooled in an electron beam cooler to equal velocities. Shown are ions for one lap in a not necessarily isochronous accelerator ring

ions of a similar mass m_b and a similar mean energy K_b will be different but it will not be completely independent of small deviations $\pm \Delta K_b$ of individual ions from the mean energy K_b. Thus there will be a small flight time addition that is proportional to $\pm \Delta K_b$ so that the accuracy of the mass measurement is slightly compromised. To avoid this deterioration of the mass measurement one can either use an energy filter to limit the velocity distribution of the ions before they are injected into the storage ring [6, 38] or determine the velocity of each individual ion by measuring its flight time [39] between two timing detectors, which are both placed in the storage ring.

There is also a second way to achieve high mass-resolving powers in an accelerator storage ring, which does not even have to be tuned to be energy isochronous. In this case a mono-energetic quasi parallel beam of electrons is passed parallel to the ion beam for some portion of the ion flight path in the accelerator storage ring. When the ions have passed many times through this region they have reached an equilibrium with the electrons and move with substantially the same speed as the electrons [40, 41] in which case the ion flight times per lap in the storage ring depend only on the ion masses. In this case (see Fig. 7) the frequency of ion passages can be recorded by Schottky pick-up electrodes similarly as in a

FTICR mass spectrograph achieving FWHM mass-resolving powers of up to and beyond $\approx$500,000 [42, 43]. However, this method can only be performed for nuclei that live longer than the ion cooling and detection process, which both require several seconds.

Using a new resonant cavity Schottky detector [44] the detection process can be sped up remarkably requiring only <1 ms to record the cycling frequency of a repeatedly passing ion. Because of this reduced recording time such a detector could even be used instead of a thin-foil secondary electron detector to record the flight time per lap of uncooled ions that move in a storage ring that is tuned to be energy isochronous [45]. In such a system possibly also incompletely cooled ions could be investigated.

As an example of such a measurement a highly resolved mass spectrum of short-lived ^{209}Bi-fragments in the rare earth region is shown in Fig. 8 as taken from reference [42]. This spectrum shows ion masses, which were partially known before and partially have been observed for the first time. Please note that in this experiment the masses of ^{143}Sm were determined in their metastable as well as in their ground state.

Please note here that mass measurements via time-of-flight techniques in storage rings are very precise and have only small errors, which at least in some cases [46] are smaller than 10keV. However, mass measurements in storage rings are also very sensitive and can determine the masses of ions that are produced with very small cross sections. This was demonstrated for instance in reference [47] in which a mass measurement was reported of ^{208}Hg of in which case only one single ion was observed during a two week beam time. Furthermore such mass measurements are also very swift so that in the most extreme case the mass of ^{133}Sb could be determined [48], which has a neutral-atom life time of only 17 μs. This makes storage ring systems uniquely suited for the investigation of short- and long-lived isomers. Please note also that with these time-of-flight techniques in large accelerator storage rings the masses of most known nuclei have been determined and that the masses of more than 30 % of them were not known before [6].

4.2 Time-of-flight mass spectrographs for low energy ions

4.2.1 Electric sector-field time-of-flight mass spectrographs

Energy isochronous time-of-flight mass analyzers can also be built from arrangements of electrostatic sector fields, provided the ion energies are only a few keV. Such energy isochronous systems are able to achieve high time-of-flight mass-resolving powers especially when built in a four-fold symmetry [49]. Similarly as magnetic systems also electrostatic ones become very powerful, when built as ring arrangements, in which a given flight path is used repeatedly. Modern systems have achieved [50, 51] FWHM mass-resolving powers $R_m \geq 100,000$. Since electrostatic sector fields can be switched quickly, it is also not too difficult to enter ions into such a time-of-flight mass analyzer or to extract them again. Disadvantageous in such systems is, however, that the sector fields and their alignment relative to each other require high mechanical accuracies, which in most cases are difficult to achieve and to maintain.

4.2.2 Mirror-type time-of-flight mass analyzers

Starting from relatively simple time-of-flight mass analyzers [52] systems were developed that use ion mirrors [24] into which ions of higher energy/charge ratios K/q penetrate deeper than ions of lower energy/charge ratios. Based on this principle energy-isochronous time-of-flight mass analyzers can be constructed as is illustrated in Fig. 9 with at least

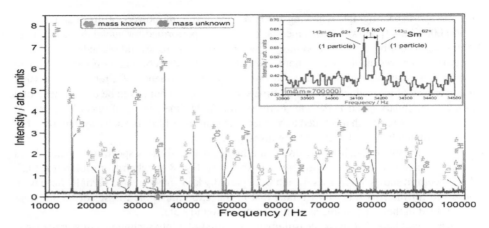

Fig. 8 A mass spectrum of short-lived nuclei in the rare earth region as recorded with the cooled beam technology showing FWHM mass-resolving powers of ≈700,000 for ions whose masses were known and of those that were observed for the first time. Please note that the mass doublet of ^{143}Sm shows the masses of this nucleus in its isomeric and in its ground state. These massed differ only by ≈5.27ppm, which is about the same as the mass difference of the mass doublet shown in Fig. 2

two different retarding field regions separated by dipole sheets which can be approximated by conductive grids. Ions of energies $K_0(1 + \delta_K)$ will pass through such an arrangement of retarding electric fields in times $T_0[1 + (T|\delta)\,\delta_K + (T|\delta\delta)\,\delta_K^2 + (T|\delta\delta\delta)\,\delta_K^3 + \cdots]$. In such systems the field strengths in the retarding field regions can be chosen [24] so that $(T|\delta) = (T|\delta\delta) = 0$ in which case the system is said to be energy isochronous to second order.

In systems that are built according to Fig. 9 the ions must pass through 5 grids before they reach the final ion detector and thus lose 5 times ≈10% of their intensity. Also the electric fields close to one of the grids are not perfect since the equipotential surfaces bulge a little through the parallel wires of a grid into the lower field region. These field penetrations cause the ions to experience actions of lenses of focal lengths $f = 2K/q(E_1 - E_2)$ when they move from a field E_1 into a field E_2, which causes distortions of the ion beam [27]. Nevertheless systems as shown in Fig. 9 have become quite successful for the mass analysis of molecular ions with mass-resolving powers of several 10,000.

As an alternative the retarding fields necessary for a time-of-flight mass spectrograph can be formed by a number of coaxial metallic rings all placed at proper potentials [53–55] as is sketched in Fig. 10. This approach requires a much more complex mathematical treatment of the system. However, at the end, grid-free ion mirrors can be built that form electric fields exactly as described by theory, provide 100% transmission, and additionally simplify the construction and use of the time-of-flight mass analyzer. So constructed time-of-flight mass spectrographs for low energy ions achieved [54–57] already in the 1980s mass resolving powers of 30,000 and of 50,000 in the 1090s.

In both gridded and grid-free retarding field time-of-flight mass spectrographs it is important to enter the ions as short pulses in Z-direction which best have been accumulated over some longer time in some storage device [29]. The length ΔT of an extracted ion pulse is here mainly determined [52, 56] by the turn-around time $\Delta T = \Delta v_z (m/qE_Z)$, i.e. twice the time $\Delta Z/\Delta v_Z$ it takes to fully decelerate an ion of mass/charge ratio m/q over a distance ΔZ in a retarding field E_Z, if the ion initially moved opposite to the direction of E_Z

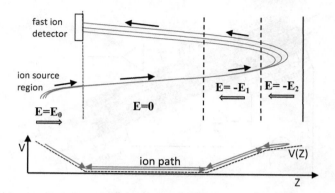

Fig. 9 Ion trajectories in a gridded energy-isochronous time-of-flight mass spectrograph ion trajectories are sketched in space and in the potential distribution formed by four regions of electric fields: E_0, 0, $-E_1$, $-E_2$. In such a system the ion flight time can be isochronous to third order, i.e. $(T|\delta) = (T|\delta\delta) = (T|\delta\delta\delta) = 0$, if the magnitudes of E_0, 0, $-E_1$, $-E_2$ are chosen properly. Note here that the grid-effects are neglected for this sketch. Note also that there are no focusing elements in this spectrograph so that the initial ion velocity in the direction perpendicular to the Z-axis persists throughout the system

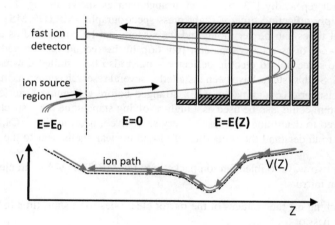

Fig. 10 Ion trajectories are sketched in space and in the potential distribution in a grid-free energy-isochronous time-of-flight mass spectrograph in which the retarding field is formed by a series of conductive rings to which appropriate potentials are applied. Also in such a system the ion flight time can be isochronous to third order, i.e. $(T|\delta) = (T|\delta\delta) = (T|\delta\delta\delta) = 0$. Please note that the first three ring electrodes form an electric round lens, which – as shown – can be used to focus the ions here for instance to a small ion detector

with a thermal velocity Δv_Z This can be achieved by cooling ions in collisions with cold neutral atoms or molecules in gases [58]. Additionally it is advantageous to form initially a low-energy ion beam that moves perpendicular to the Z-axis and that is widened by proper lens arrangements so that the unavoidable angles of inclination of the ion trajectories and thus the ions' velocity components Δv_Z are reduced [59, 60]. However, there is also a disadvantage coupled with this method in that the ions intensity is decreased, if the accepted diameter of the ion beam is limited.

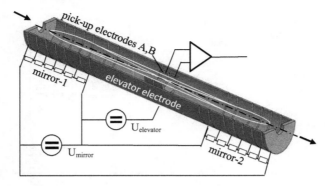

Fig. 11 An energy-isochronous multi-reflection time-of-flight mass spectrograph. In this system ions are reflected repeatedly between two focusing ion mirrors each of which consists of 6 ring electrodes and one end cap. The ion trajectories - shown in white - illustrate a point-parallel-point focusing of the ions for each lap, i.e. the focal lengths of the ion mirrors are $1/2$ of one lap or which is the same $1/2$ of the length of the MRTOF-MS

Multi-reflection time-of-flight mass spectrographs The use of grid-free ion mirrors in energy-isochronous time-of-flight mass spectrographs opens the possibility to use the ion flight path repeatedly [53, 61] in an arrangement as shown in Fig. 11, i.e. in a so-called "multiply-reflecting time-of-flight mass spectrograph" (MRTOF MS). In this case the ion flight times can be very long and the system can be configured as a mass spectrograph [61–64] or as a mass separator [65, 66]. In the second case an additional beam shutter – like some pulsed electric deflector – must also be installed downstream of the MRTOF-MS. Such systems have been installed at several research centers as mass prefilters that remove isobaric contaminants, when arranged before a Penning Trap [5]. However, the same system could also be used as a high-resolving standalone mass spectrograph [67] and so allowed to determine the mass of very short-lived, neutron rich ^{52}Ca, whose mass is important to understand the properties of closed nuclear shells near to the neutron drip line.

There are two ways to inject an ion packet into the MR-TOF MA or to eject it through one of the ion mirrors:

1. By switching off the voltages of the mirror electrodes for a short time [61–65] so that ions can pass or
2. By increasing the energy of the ions, when passing through a mirror and having a tubular elevator electrode at a correspondingly high potential, which is reduced, when the ion packet is inside this elevator electrode [66, 67]

Since in all cases the ion injection time is very short as compared to the ion flight time between the two mirrors it is always advantageous to accumulate ions for a certain time in a storage ion source [29, 61–63] or in a RF-ion trap [64] before they are injected into the ion race-track between the ion mirrors.

There are also two ways to precisely determine the overall ion flight times in the MRTOF-MS, when the ion mirrors are tuned so that energy isochronicity per lap is achieved. It is possible:

1. To Fourier transform the signals induced by the passing ions in the azimuthally split pick-up electrodes A and B and so determine the passage frequencies characteristic of ions of different masses [68].

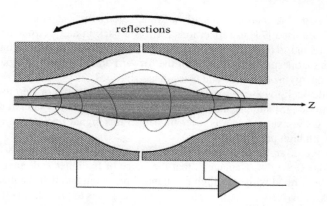

Fig. 12 Sketch of an Orbitrap and some trajectories between the inner and outer coaxial electrodes. Note here that the ions are reflected energy-isochronously between the two ion mirrors that are formed, when the axial electric field increases beyond some limit at position at which the distance between the inner and outer electrode reduce. Note also that the potential differences induced in the two sections of the outer electrodes are being amplified as the signal that must be Fourier transformed to provide the frequencies of ion passage. Note finally that the shown ion trajectories are not calculated but shown for general illustration purposes only

2.   To switch off one of the ion mirrors after ions of a chosen mass have performed N laps in the ion-race-track of the MRTOF MS and allow them to impinge on a sensitive fast ion detector [61–64] outside ion-race-track.

In case (1) the recording may take a long time, but an overall mass spectrum is obtained. However, its signal strength will be small in most cases. In case (2) ions are recorded with a high efficiency after they have performed many laps. However, there are ions of different sections of the mass spectrum superimposed as caused by ions that have performed different number of laps at the time of ion extraction. When on the other hand ions of a chosen mass are recorded in different time recordings after they have performed N and N+n laps with $n \ll N$, the overall mass spectrum can be reconstructed [69]. Using both methods (1) and (2) in close sequence the interpretation of the obtained mass spectra can assist each other in that occasionally occurring ambiguities can be eliminated.

For precise mass measurements precise reference masses must be provided, like ^{12}C-clusters [5] or chosen molecular ions [22]. The standard procedure here is to use reference mass ions of which at least one is slightly heavier and one is slight lighter than the ion whose mass is to be determined. However, in the case of a time-of-flight mass spectro-graph one single reference mass suffices [70], since the mass dispersion of the system is determined by a smooth mathematical function as long as the supply voltages are constant.

Remarkable for all MRTOF MS is that the achievable mass-resolving power is high and that the system is very robust and can work in a harsh environment as had been required and achieved for instance for space missions [71, 72].

The orbitrap Forming two rotationally symmetric surfaces and arranging them coaxially around a Z-axis (see Fig. 12) and applying to them different potentials, properly injected ions can rotate around the inner surface and are still free to move along the Z-axis [73]. Shaping the electrodes as shown in Fig. 12 the potential along the Z-axis can be arranged to change with Z^2, in which case two electrostatic mirrors are formed between which an axial back-driving field increases linearly with ±Z along the Z-axis as in ref. [57]. Such mirrors have been used [74] to reflect ions forth and back in an electrode arrangement as shown in Fig. 12 and so form an ion trap in which ions still rotate around the inner electrode. Remarkable is that despite of this rotation the forward and backward motion between the ion mirrors is energy-isochronous and mass-dispersive [75].

Since it is difficult to extract ions from such a trap, it was a major step forward to split the outer electrode [75] and use the induced signal between the two parts to record the ion motion along the Z-axis. After a Fourier transform this signal reveals the masses of the moving ions with high precision.

The advantage of an Orbitrap is its remarkably high mass-resolving power [75–77] of more than 100,000. Its disadvantage is that the shape of the inner and outer electrodes of the trap must be formed with very high mechanical precision and that the mass analysis of a sample takes a relatively long time. Thus it is advantageous to provide an efficient storage of ions for instance in a Paul trap between consecutive mass measurements.

5 Conclusion

Mass analyzers of high resolving power are being built in many different ways. The highest demands are so far met by Penning traps and accelerator storage rings for the mass analysis of atomic ions and here especially those of short-lived nuclei because of their importance for astrophysics. Increasingly high demands arise also in the mass analysis of complex molecules for biological, pharmaceutical and environmental research. In recent years many instruments have been developed by industry for these applications. Rather wide acceptance can be seen for Orbitraps and nondestructive Penning traps. Also it can be expected that more and more multi-reflection time-of-flight mass analyzers will be employed, because of their relatively low demands on mechanical alignments and because of their ease of handling.

References

1. Aston, F.W.: Philos. Mag. **38**, 709 (1919)
2. Mattauch, J., Herzog, R.: Zeitschrift für Physik **89**, 786 (1934)
3. Nier, A.O.: Phys. Rev. **50**, 1041 (1936)
4. Wollnik, H.: Optics of Charged Particles. Academic Press, Orlando (1987)
5. Blaum, K.: Phys. Rep. **425**, 1 (2006)
6. Bosch, F., Litvinov, Yu., Stoehlker, T.: Prog. Part. Nucl. Phys. **73**, 84 (2013)
7. Irnich, H., Geissel, H., et al.: Phys. Rev. **75**, 4182 (1995)
8. Wanjo, S., Goriely, S., et al. Astrophys. J. **606**, 1057 (2004)
9. Russel, D.H., Edmonson, R.D.: J. Mass Spectrom. **32**, 263 (1997)
10. Hopfgartner, G., Tonelli, D., Varesio, E.: Anal. Bioanal. Chem. **402**, 2581 (2012)
11. Ewald, H.: Z. Naturforsch. **1**, 131 (1946)
12. Fukumoto, S., Matsuo, T., Matsuda, H.: J. Phys. Soc. Jpn. **25**, 946 (1968)
13. Wollnik, H., Ewald, H.: Nucl. Instr. Methods **36**, 93 (1965)
14. Marshall, A.G., Hendrickson, C.L.: Mass Spectrom. Rev. **88**, 319 (1998)
15. Boldin, I.A., Nikolaev, E.N.: Rapid Commun. Mass Spectrom. **25**, 122 (2011)
16. Nikolaev, E.N.: 65th ASMS meeting Proc. Baltimore (2014)
17. Marshall, A.G., Rodgers, R.: Proc. National Acad. Sci. **105**, 18090 (2008)
18. Vladimirov, G., Hendrickson, C.L., et al.: J. Am. Soc. Mass Spectrom. **23**, 375 (2012)
19. Bollen, G., Moore, R.B., Savard, G., Stolzenberg, H.: J. Appl. Phys. **68**, 4355 (1990)

20. Kluge, H.J., Bollen, G.: Nucl. Instr. Methods B **70**, 473 (1992)
21. Eliseev, S., Blaum, K., et al.: Phys. Rev. Lett. **110**, 082501 (2013)
22. Naimi, S., Nakamura, S., et al.: J. Mass Spectrom. Ion Phys. **340**, 38 (2013)
23. Poschenrieder, W.: Int. J. Mass Spectrom. Ion Phys. **9**, 357 (1972)
24. Mamyrin, B.A., Karateev, V.I.: Sov. Phys. JETP **37**, 45 (1973)
25. Wollnik, H., Matsuo, T.: Int. J. Mass Spectrom. Ion Phys. **37**, 209 (1981)
26. Wollnik, H.: Nucl. Inst. Methods **186**, 441 (1981)
27. Wollnik, H.: Int. J. Mass Spectrom. Ion Phys. **349**, 38 (2013)
28. Kraus, R., Vieira, D.J., Wollnik, H., Wouters, J.M.: Nucl. Inst. Methods A **264**, 327 (1988)
29. Grix, R., Grüner, U., et al.: Int. J. Mass Spectrom. Ion Phys. **93**, 323 (1989)
30. Wouters, J.M., Vieira, D.J., Wollnik, H., et al.: Nucl. Inst. Methods A **240**, 77 (1985)
31. Wouters, J.M., Vieira, D.J., Wollnik, H., et al.: Nucl. Inst. Methods B **26**, 286 (1987)
32. Wollnik, H., Schwab, T., Berz, M.: GSI Rep. **86-1**, 372 (1986)
33. Wollnik, H.: Nucl. Inst. Methods B **26**, 267 (1987) **A258**, 289 (1987)
34. Troetscher, J., Balog, K., et al.: Nucl. Inst. Methods B **70**, 455 (1992)
35. Klein, Ch., Troetscher, J., Wollnik, H.: Nucl. Inst. Methods A **335**, 146 (1993)
36. Wollnik, H., Beckert, K., et al.: Nucl. Phys. **626**, 327 (1996)
37. Stadlmann, J., Hausmann, M., et al.: Phys. Lett. B **586**, 27 (2004)
38. Geissel, H., Knoebel, R., et al.: Hyperfine Interact. **173**, 49 (2006)
39. Zhang, Y., Tu, X.L., et al.: Nucl. Instr. Methods A **756**, 1 (2014)
40. Franzke, B.: Nucl. Instr. Methods B **24**, 18 (1987)
41. Radon, T., Geissel, H., et al.: Nucl. Phys. A **677**, 75 (2000)
42. Attallah, F., Hausmann, M., et al.: Nucl. Phys. A **701**, 561 (2002)
43. Litvinov, Y., Geissel, H., et al.: Nucl. Phys. A **759**, 23 (2005)
44. Nolden, F., Huelsmann, P., et al.: Nucl. Phys. A **701**, 56 (2002)
45. Walker, F., Litvinov, Yu., Geissel, H.: Int. J. Mass Spectrom **349**, 247 (2013)
46. Tu, X.L., Wang, M., et al. Nucl. Instr. Methods A **654**, 213 (2011)
47. Chen, I., Yu, L., et al.: Phys. Rev. Lett. **102**, 122503 (2000)
48. Sun, B., Knoebel, R., et al.: Phys. Lett. B **688**, 294 (2010)
49. Sakurai, T., Matsuo, T., Matsuda, H.: Int. J. Mass Spectrom. Ion Phys. **63**, 273 (1985)
50. Nishiguchi, M., Ueno, Y., Toyoda, M., Setou, M.: J. Mass Spectrom. **44**, 594 (2009)
51. Satoh, T., Sato, T., Kubo, A., Tamura, J.: J. Am. Soc. Mass Spectrom. **22**, 797 (2011)
52. Wiley, W.C., McLaren, I.H.: Rev. Sci. Instrum. **26**, 1150 (1955)
53. Wollnik, H.: Patent DE3025764 (1982)
54. Wollnik, H., Grix, R., et al.: Proc. 2. Japan-China Symp. on Mass Spectrometry, p. 181. Bando Press, Osaka (1987)
55. Kutscher, R., Grix, R., Li, G., Wollnik, H.: Int. J. Mass Spectrom. Ion Phys. **103**, 117 (1991)
56. Wollnik, H.: Int. J. Mass Spectrom. Ion Phys. **131**, 387 (1994)
57. Cotter, R., Doroshenko, W.: Patent US6.365.892 B1 (2002)
58. Kozlovsky, V., Fuhrer, K., et al.: Int. J. Mass Spectrom. Ion Phys. **181**, 27 (1998)
59. Dodonov, A.F., Chernushevich, I.V., et al.: USSR patent 1681340A1 (1987)
60. Dodonov, A.F., Kozlovsky, V.I., et al.: Eur. J. Mass Spectrom. **6**, 481 (2000)
61. Wollnik, H., Przewloka, M.: Int. J. Mass Spectrom. Ion Phys. **96**, 267 (1990)
62. Wollnik, H., Casares, A.: Int. J. Mass Spectrom. Ion Phys. **227**, 217 (2003)
63. Ishida, Y., Wada, M., Wollnik, H.: Nucl. Instr. Methods B **21**, 241 (2005)
64. Schury, P., Wada, M., et al.: Nucl. Instr. Methods B **335**, 39 (2014)
65. Piechaczek, A., Shchepunov, V., et al.: Nucl. Instr. Methods B **266**, 4510 (2008)
66. Wolf, R.N., Beck, D., Blaum, K., et al.: Nucl. Instr. Methods **686**, 82 (2012)
67. Wienholtz, F., Beck, D., et al.: Nature **498**, 346 (2013)
68. Radford, D., Casares, A.: Private communication (2003)
69. Schury, H., Ito, Y., Wada, M., Wollnik, H.: Int. J. Mass Spectrom. Ion Phys. **359**, 19 (2014)
70. Ito, Y., Schury, P., et al.: Phys. Rev. C **88**, 11306 (2011)
71. Casares, A., Kholomeev, A., et al.: 47th ASMS meeting Proc. Dallas (1999)
72. Balsiger, H., et al.: Adv. Space Res. **21**, 1527 (1998)
73. Kingdon, K.H.: Phys. Rev. **21**, 408 (1923)
74. Knight, R.D.: Appl. Phys. Lett. **38**, 4 (1981)
75. Gall, L.N., Golikov, Y.K.: USSR patent 1247973 (1986)
76. Makarov, A.A.: Anal. Chem. **72**, 2113 (2000) **78**, 1156 (2006)
77. Makarov, A.A., Grinfeld, D.E., Monastrilsky, M.A.: Nauchnoe Priborostroebnie **24**, 68 (2014)

  Springer

Hyperfine Interact (2015) 235:77–86
DOI 10.1007/s10751-015-1187-z

Status of the TAMUTRAP facility and initial characterization of the RFQ cooler/buncher

M. Mehlman[1] · P. D. Shidling[1] · R. Burch[1] · E. Bennett[1] ·
B. Fenker[1] · D. Melconian[1]

Published online: 13 May 2015
© Springer International Publishing Switzerland 2015

Abstract The Texas A&M University Penning Trap experiment (TAMUTRAP) is an upcoming ion trap facility that will be used to search for possible scalar currents in $T = 2$ superallowed pure Fermi decays (utilizing β-delayed proton emitters), which, if found, would be an indication of physics beyond the standard model. In addition, TAMUTRAP will provide a low-energy, point-like source of ions for various other applications at the Cyclotron Institute at Texas A&M University. The experiment is centered around a unique, compensated cylindrical Penning trap that employs a specially optimized length/radius ratio in the electrode structure that is not used by any other facility. The radioactive beam, provided by the Texas A&M University Re-accelerated Exotics (T-REX) program at the Cyclotron Institute, will be prepared for loading in the measurement trap via a specially designed Radio Frequency Quadrupole (RFQ) cooler/buncher. These proceedings will describe the current status of the TAMUTRAP facility, paying particular attention to the design and initial characterization of the RFQ cooler/buncher. Future plans will also be discussed.

Keywords Penning · Precision · RFQ · Resign · Cooler · Buncher · Status

1 Introduction

Low energy precision β-decay studies have proven to be an excellent complement to high-energy physics experiments for placing new constraints on physics beyond the standard model (SM) [2, 7, 8]. Up to this point, it has been possible to explain the results from such

Proceedings of the 6th International Conference on Trapped Charged Particles and Fundamental Physics (TCP 2014), Takamatsu, Japan, 1-5 December 2014

✉ M. Mehlman
mehlmanmichael@tamu.edu

[1] Cyclotron Institute, Texas A&M University, 3366 TAMU, College Station, TX 77843-3366, USA

Table 1 The $T = 2$ nuclei that will compose the initial experimental program measuring $a_{\beta\nu}$

Nuclide	Lifetime (ms)	E_p (MeV)	R_L (mm)
^{20}Mg	131.02	4.28	42.7
^{24}Si	202.74	3.91	40.8
^{28}S	180.37	3.70	39.7
^{32}Ar	145.00	3.36	37.8
^{36}Ca	147.19	2.55	33.0
^{40}Ti	75.60	3.73	39.9
^{48}Fe	65.36	1.23	22.9

The Larmour radii, R_L, for the ejected protons of interest (having energy E_p) shown are calculated for the 7T magnetic field at TAMUTRAP

experiments by a time reversal invariant $V - A$ interaction displaying a maximal violation of parity; however, more precise measurements of ft values [9, 10] and correlation parameters [1] in particular β-decays can serve to test for the presence and properties of any non-SM processes that may occurr in such interactions.

1.1 Motivation

The initial experimental program at TAMUTRAP will seek to improve the limits on non-SM processes in the weak interaction, in particular scalar currents, by measuring the $\beta - \nu$ correlation parameter, $a_{\beta\nu}$, for $T = 2, 0^+ \rightarrow 0^+$ superallowed β-delayed proton emitters (the preliminary list of nuclei to be studied is outlined in Table 1).

The general β-decay rate with no net polarization or alignment is given by [4]:

$$\frac{d^5\Gamma}{dE_e d\Omega_e d\Omega_\nu} \propto 1 + \frac{p_e}{E_e}a_{\beta\nu}\cos\theta_{\beta\nu} + b\frac{m_e}{E_e}, \tag{1}$$

where, E_e, p_e, and m_e are the energy, momentum, and mass of the β, $\theta_{\beta\nu}$ is the angle between the β and ν, and b is the Fierz interference coefficient. Thus, it is possible to determine the $\beta - \nu$ correlation parameter by means of an experimental measurement of the angular distribution between the β and ν. For the strict $V - A$ formulation of the weak interaction currently predicted by the SM, the $\beta - \nu$ angular distribution in a pure Fermi decay should yield a value for $a_{\beta\nu}$ of exactly 1. Any admixture of a scalar current to the predicted interaction, a result of particles other than the expected $W^\pm$ being exchanged during the decay, would result in a measured value of $a_{\beta\nu} < 1$.

TAMUTRAP will observe this angular distribution between β and ν for β-delayed proton emitters. In such a case, the β-decay yields a daughter nucleus that is unstable, and can result in the subsequent emission of a proton with significant probability. As discussed in [1], the great advantage to utilizing β-delayed proton emitters for such a study is that this proton energy distribution contains information about $\theta_{\beta\nu}$. If the β and ν are ejected from the parent nucleus in the same direction, they will impart a larger momentum kick to the daughter nucleus, which will be inherited by the proton. Conversely, if the β and ν are emitted in opposite directions, this momentum kick is reduced. By measuring the proton energy distribution at TAMUTRAP the value of $a_{\beta\nu}$ will be deduced, which can then indicate the existence of a scalar current in these decays [4].

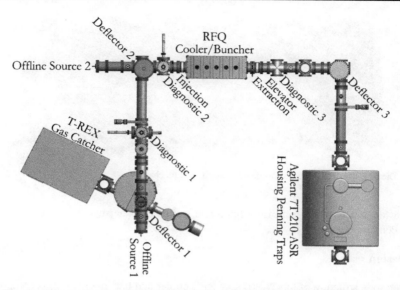

Fig. 1 The planned beamline for the TAMUTRAP facility. The T-REX gas catcher, part of the Heavy Ion Guide, is located below plane of the TAMUTRAP setup, and is connected via a straight section of vertical beamline. Major components are annotated

1.2 Experimental approach

Such precision β-decay experiments are well served by a cylindrical Penning trap due to the fact that the magnetic field employed to confine the ions radially may simultaneously be used to contain charged decay products [5] such as β's and protons with up to 4π acceptance in an appropriately designed trap [6]. The MeV-energy protons of interest (as well as less magnetically rigid betas) easily escape the shallow axial potential well, and are guided along helical paths toward the endcaps of the trap by the strong magnetic radial confinement. Position sensitive silicon detectors are planned for proton and beta detection at these endcaps, yielding near complete geometric efficiency. At the same time, features of a cylindrical trap geometry can be useful for other nuclear physics experiments, such as maintaining a line of sight to the trap center for spectroscopy, an easily "tunable" and "orthogonalized" electric field for experiments requiring a harmonic potential (such as mass measurements), and unrivaled access to the trapped ions. The planned TAMUTRAP beamline is shown in Fig. 1.

2 RFQ cooler/buncher

In order to efficiently load ions into a Penning trap, the beam should be bunched and have a low energy with sufficiently small time and energy spread, *i.e.* low emittance. A gas-filled linear RFQ Paul trap is particularly adept at such beam preparation, and has been developed for use at TAMUTRAP. The ions of interest are cooled through collisions with a low-mass buffer gas, typically Helium [3], reducing the phase space of the beam. After a sufficient number of collisions, the ions collect around the bottom of the potential well and are thermalized to a largely uniform energy (and spatial) distribution. The resulting bunch can be ejected by switching off the axial potential well and is sent toward the measurement

Springer

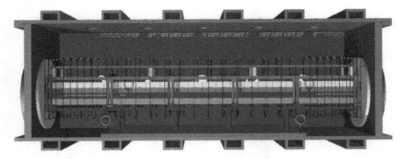

Fig. 2 The RFQ structure, situated inside the custom vacuum chamber

trap for loading with a significantly improved emittance better matched to the acceptance of the Penning trap.

2.1 Design

The electrode structure of the TAMUTRAP RFQ cooler and buncher (Fig. 2) is composed of four rods with radius of curvature $r = 7$ mm that are rigidly held at a surface-to-surface rod spacing of $2r_0 = 12$ mm for opposite rods, yielding a characteristic ratio of $r/r_0 = 1.1\bar{6}$. The structure is approximately 87 cm in length, and is separated axially into 33 segments to enable the application of a linear drag potential. Segment lengths are 38.1 mm for the majority of the device. However, 19.1 mm and 9.6 mm segments are employed at each end to allow finer control of axial DC potentials at these locations, which is particularly useful for optimizing injection and extraction, and also allows for forming a tighter ion bunch. Electrically isolated entrance and extraction electrodes are situated at each end of the device to allow for one final means of tuning the ion beam in order to ensure the greatest acceptance and smallest bunch dimensions on extraction. The entire structure is completely symmetric about the axial mid-plane to allow both forward and reverse operation, adding yet another element of flexibility to the TAMUTRAP facility.

The mechanical structure for this device has been optimized to ensure mechanical rigidity, hide dielectrics, and achieve the minimum gap between adjacent segments. Care has also been taken to minimize electrical impedance of the RF structures by minimizing material in critical locations. All electrode spacings are held at 0.38±0.13 mm, and are toleranced such that the total length of the electrode structure can vary by no more than ±0.13 mm as well. Transverse movement of electrodes is similarly limited to less than 0.13 mm, and angular mis-alignment is only tolerated by typically 0.2°. The main structure is composed of only 8 custom fabricated precision parts, with the remainder of the assembly coming from precision stock components. Apart from electronics, all components used are made of aluminum, stainless steel, or ceramic for vacuum considerations. A photo of the assembled device can be seen in Fig. 3.

Analog electronics have been developed to drive the device, with each segment receiving a unique adjustable DC potential for fine-tuning of the axial electric field. RF is coupled to the segments in vacuum using vacuum safe ceramic capacitors and resistors, ensuring a minimum of line-impedance. Switching of the final segments during ejection is accomplished by a single Behlke HTS 31-03-GSM high voltage, ultra fast solid-state switch. The switch itself demonstrates a switching time on the order of 500 ns, which is slowed to approximately 50 μs due to the RC circuit attached to each electrode that is used to protect the

Fig. 3 The RFQ structure with in-vacuum RF and DC circuitry during test-assembly

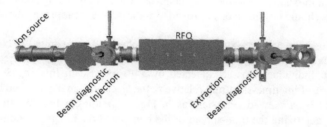

Fig. 4 The beam line used for characterization of the RFQ cooler/buncher

DC power supply. Despite the relatively slow switching time, satisfactory bunch characteristics have been observed, as will be discussed. A digital square-wave RF generator is being considered as a replacement for the analog driver after discussion initiated at TCP 2014 (P. Schury, personal communication, December 2014).

2.2 Preliminary characterization

The cooler/buncher device has been assembled, and commissioning has begun as of November 2014. Initial characterization of the device includes performing efficiency measurements in continuous mode, and describing the properties of ion bunches that result from operating the device in bunched mode.

Continuous mode efficiencies are calculated as the ratio of the beam current measured on Faraday cups located prior to the injection optics and after the extraction optics of the cooler/buncher, as indicated in Fig. 4. As such, the efficiency of some additional ion optics are taken into account in these tests, and the results should be taken as a lower limit of the pure RFQ efficiency. Preliminary results are shown in Fig. 5, and all efficiencies are reported at 0 V/mm drag potential. The efficiency for each energy and pressure combination was found to be optimized by a distinct drag potential setting, so the decision was made to facilitate comparison by choosing a constant 0 V/mm rather than possibly introducing

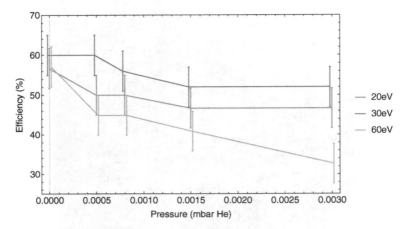

Fig. 5 Continuous-mode efficiencies as a function of gas pressure (all reported with a drag potential of 0 V/mm)

additional error into the measurements by choosing a sub-optimal drag potential. As a result, all efficiencies should be able to be improved to some degree by adjusting the drag potential.

In normal use, the TAMUTRAP RFQ cooler/buncher will be operated in bunched mode, collecting ions of interest for some set amount of time, bunching, and ejecting them in a tight packet. Individual ions are detected by a 40 mm Beam Imaging Solutions MCP detector. The resulting time-spectrum relative to the ejection signal generated by the control system was fit by a skewed Gaussian, as in Fig. 6, yielding a Full Width at Half Max (FWHM) characterizing the time-spread of the bunch and the integrated number of counts per bunch (up to an arbitrary constant dependent on acquisition and analysis). It should be noted when comparing bunch characteristics to other facilities that fitting with a standard Gaussian resulted in a poorer fit, but also significantly reduced the observed FWHM due to exclusion of the low-count large-time tail of the distribution. Caution should be observed when comparing integrated number of counts between data points, as fluctuation of up to 10% in ion source current was observed on a several minute time scale.

The operation of the RFQ in bunched mode was investigated systematically by testing the effect of adjusting one operation parameter at a time. While it is true that various parameters are no-doubt correlated, the parameter space was too large to evaluate the variables co-dependently. A small subset of the operating parameters tested systematically is presented here.

At the pressures available for operation at TAMUTRAP ($< 5 \times 10^{-3}$ mbar He), the ultimate bunch characteristics have proven to be largely independent of gas-pressure. At the low-pressure extreme, the integrated number of counts begins to fall off, since there is a minimum amount of buffer-gas required for successfully cooling and bunching the incoming ions. This makes no comment on the effect gas pressure has on transverse emittance, which could worsen to some degree with increasing pressure due to gas collisions after ejection. FWHM and number of ions per bunch as a function of gas pressure can be seen in Fig. 7.

A 30 eV incident beam energy demonstrated the greatest continuous mode transmission efficiency of all incident energies tested for an uncooled beam. Since the TAMUTRAP RFQ will be employed exclusively as a cooler/buncher, it is more critical to determine what beam

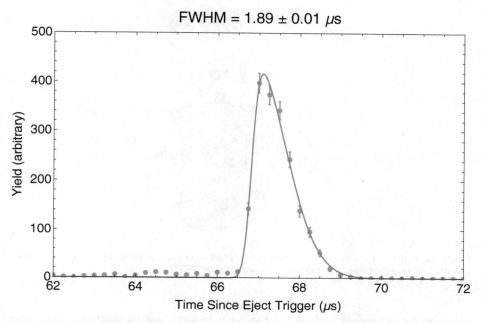

Fig. 6 A good bunch with skewed Gaussian fit superimposed. Beam energy = 30 eV, gas pressure = 3 × 10^{-3} mbar He, drag potential = 0.08 V/cm, incident beam current = 0.7 pA. FWHM of Gaussian fit = 1.89 μs

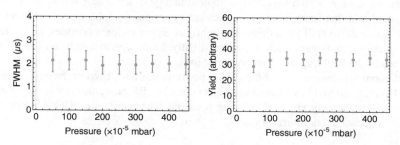

Fig. 7 Bunch FWHM (*left*) and yield (*right*) as a function of buffer gas pressure. Beam energy = 30 eV, drag potential = 0.08 V/cm. Bunches did not form below about 25 mbar He

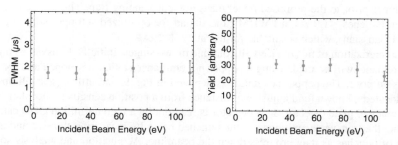

Fig. 8 Bunch FWHM (*left*) and yield (*right*) as a function of incident beam energy. Gas pressure = 3 × 10^{-3} mbar He, drag potential = 0.08 V/cm

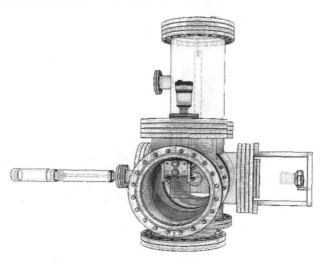

Fig. 9 The design of the pepper pot emittance station which has been constructed for transverse emittance characterization of the cooled and bunched beam output by the RFQ

energy to use in bunched mode in order to obtain bunches with the smallest FWHM time spread and greatest yield (Fig. 8). This was accomplished by raising and lowering the voltage at which the RFQ platform is floated in order to achieve the desired potential difference from the ion source, which was held at approximately 10 kV. The FWHM of the bunch's time spread is rather insensitive to the incident beam energy (phase space is reset in the device), while the overall yield degrades slightly at higher incident energies. In this regime, the number of counts per bunch decreases slightly, likely due to a reduced initial capture efficiency of the Paul trap for more energetic ions.

Additional systematic tests of the device performance have been performed, measuring bunch FWHM and yield as a function of eject duration, RF properties (frequency and voltage), incident beam current, and DC drag potential. These tests are still being concluded, and will be presented in future work.

2.3 Outlook

The TAMUTRAP cooler/buncher has undergone initial characterization for use as a beam conditioner prior to the proposed novel measurement Penning trap. The device has generated ion bunches ≤ 1.9 μs in FWHM time spread for optimized settings (when fit with a skewed Gaussian), which is suitable for use at TAMUTRAP.

Characterization of this device will continue in two stages. Initially, transverse emittance measurements will be made using a pepper pot emittance station developed at TAMUTRAP for this purpose. The pepper pot station is depicted in Fig. 9, and utilizes a small upstream mask located a known (and adjustable) distance from a position sensitive, phosphor-backed micro channel plate detector. This detector is read out by a high definition CCD camera to calculate the transverse emittance of the bunched beam by measuring the size and angular spread of bunches as they progress down the beamline. Acquisition and analysis software has been developed for this system, and has been initially validated offline using simple ray-tracing simulations.

Subsequently, a measurement will be made of the energy spread of the resulting bunch in order to obtain an understanding of the longitudinal emittance. This project will require the development of additional experimental apparatus needed to measure the energy spectrum of the resulting ions.

With the commissioning of the RFQ cooler/buncher, the entire TAMUTRAP facility is progressing closer to completion. Upcoming tasks will include continuing measurement trap and detection simulations with GEANT4, mechanical and electrical design of the measurement traps and detection systems, fabrication of these components, coupling of the measurement system to the TAMUTRAP beamline, and, finally, commissioning of the facility and first measurements. Time lines for these tasks are tentative; however, GEANT simulations should be completed by the end of 2015, with mechanical design and fabrication occurring in 2016. Commissioning and first use of the facility is currently estimated for the year 2017.

3 Conclusion

A new RFQ cooler/buncher has been constructed to feed a novel measurement Penning trap designed with the initial research program of measuring correlation parameters for $T = 2$ superallowed β-delayed proton emitters in mind. Careful attention has been paid to create a device with maximum suitability for a wide range of possible future nuclear physics experiments. In particular, a symmetric, finely tunable, and mechanically rigid electrode structure will allow the cooler/buncher to be operated in forward and reverse modes for maximum flexibility (though the currently proposed experimental program only calls for operation in the forward mode). This device has been constructed, and a preliminary characterizations has been performed. Complete characterization should be concluded in 2015, allowing for further development of the TAMUTRAP facility in 2015 and beyond.

Acknowledgments   The authors would like to thank Guy Savard, Jason Clark, Ryan Ringle, and Jens Dilling for fruitful discussions and advice. This work was supported by the U.S. Department of Energy under Grant Numbers ER41747 and ER40773.

Conflict of interests   The authors declare that they have no conflicts of interest.

References

1. Adelberger, E.G., Ortiz, C., Garcia, A., Swanson, H.E., Beck, M., Tengblad, O., Borge, M.J.G., Martel, I., Bichsel, H.: Positron-neutrino correlation in the $0^+ \rightarrow 0^+$ Decay of ^{32}Ar. Phys. Rev. Lett. **83**(7), 1299–1302 (1999)
2. Behr, J.A., Gwinner, G.: Standard model tests with trapped radioactive atoms. J. Phys. G: Nucl. Part. Phys. **36**(3), 033101 (2009)
3. Herfurth, F., Dilling, J., Kellerbauer, A., Bollen, G., Henry, S., Kluge, H.J., Lamour, E., Lunney, D., Moore R.B., Scheidenberger, C., et al.: A linear radiofrequency ion trap for accumulation, bunching, and emittance improvement of radioactive ion beams. Nucl. Instrum. Methods Phys. Res., Sect. A **469**(2), 254–275 (2001)
4. Jackson, J.D., Treiman, S.B., Wyld, H.W.: Coulomb corrections in allowed beta transitions. Nucl. Phys. **4**, 206–212 (1957)
5. Kozlov, V.Y., Beck, M., Coeck, S., Herbane, M., Kraev, I.S., Severijns, N., Wauters, F., Delahaye, P., Herlert, A., Wenander, F., Zakoucky, D.: The WITCH experiment: Towards weak interactions studies. Status and prospects. Hyperfine Interact. **175**, 15–22 (2006)

6. Mehlman, M., Shidling, P.D., Behling, S., Clark, L.G., Fenker, B., Melconian, D.: Design of a unique open-geometry cylindrical penning trap. Nucl. Instrum. Methods Phys. Res., Sect. A **712**, 9–14 (2013)
7. Severijns, N., Beck, M., Naviliat-Cuncic, O.: Tests of the standard electroweak model in beta decay. Rev. Mod. Phys. **78**, 991–1040 (2006)
8. Severijns, N., Naviliat-Cuncic, O.: Symmetry tests in nuclear beta decay. Annual Review of Nuclear and Particle Science **61**, 23–46 (2011)
9. Severijns, N., Tandecki, M., Phalet, T., Towner, I.S.: Ft values of the T=1/2 mirror beta transitions. Phys. Rev. C **78**, 055501 (2002)
10. Towner, I.S., Hardy, J.C.: The evaluation of $V_u d$ and its impact on the unitarity of the Cabibbo Kobayashi Maskawa quark-mixing matrix. Rep. Prog. Phys. **73**(4), 046301 (2010)

Hyperfine Interact (2015) 235:87–95
DOI 10.1007/s10751-015-1192-2

Using GPU parallelization to perform realistic simulations of the LPCTrap experiments

X. Fabian[1] · F. Mauger[1] · G. Quéméner[1] · Ph. Velten[2] · G. Ban[1] · C. Couratin[1] ·
P. Delahaye[3] · D. Durand[1] · B. Fabre[4] · P. Finlay[2] · X. Fléchard[1] · E. Liénard[1] ·
A. Méry[5] · O. Naviliat-Cuncic[6] · B. Pons[4] · T. Porobic[2] · N. Severijns[2] ·
J. C. Thomas[3]

Published online: 11 June 2015

Abstract The LPCTrap setup is a sensitive tool to measure the $\beta - \nu$ angular correlation
coefficient, $a_{\beta\nu}$, which can yield the mixing ratio ρ of a β decay transition. The latter
enables the extraction of the Cabibbo-Kobayashi-Maskawa (CKM) matrix element V_{ud}. In
such a measurement, the most relevant observable is the energy distribution of the recoiling
daughter nuclei following the nuclear β decay, which is obtained using a time-of-flight
technique. In order to maximize the precision, one can reduce the systematic errors through
a thorough simulation of the whole set-up, especially with a correct model of the trapped ion
cloud. This paper presents such a simulation package and focuses on the ion cloud features;
particular attention is therefore paid to realistic descriptions of trapping field dynamics,
buffer gas cooling and the N-body space charge effects.

Keywords Paul trap · LPCTrap · GPU · Parallelization · Simulation · N-body

Proceedings of the 6th International Conference on Trapped Charged Particles and Fundamental
Physics (TCP 2014), Takamatsu, Japan, 1–5 December 2014.

✉ X. Fabian
   fabian@lpccaen.in2p3.fr

[1] LPC-Caen, ENSICAEN, Université de Caen, CNRS/IN2P3, Caen, France

[2] KU Leuven, Instituut voor Kern- en Straglingsfysica, Leuven, Belgium

[3] GANIL, CEA/DSM-CNRS/IN2P3, Caen, France

[4] CELIA, Université de Bordeaux, CEA/CNRS, Bordeaux, France

[5] CIMAP, CEA/CNRS/ENSICAEN, Université de Caen, Caen, France

[6] NSCL and Department of Physics and Astronomy, MSU, East-Lansing, MI 48824, USA

1 Context & motivations

The precise measurement of the β-ν angular correlation coefficient, $a_{\beta\nu}$, in nuclear β decay is a sensitive tool to search for exotic couplings presently excluded by the V-A theory of the Standard Model. In the case of a mixed mirror transition, $a_{\beta\nu}$ also allows one to determine the mixing ratio ρ. In that case, the measurement of $a_{\beta\nu}$ constitutes an important input for the database of nuclear mirror transitions, enabling the extraction of the CKM matrix element V_{ud} with improved precision for such transitions [1]. In the LPCTrap device, the radioactive nuclei are confined in a Paul trap, allowing the detection of the recoil ions in coincidence with the β particles [2]. This time-of-flight (TOF) technique allows to determine with an excellent precision not only $a_{\beta\nu}$, but also the shake-off probabilities inherent in the decay of singly charged ions [3, 4]. Several measurement campaigns were carried out for the last ten years for three different nuclei (^{6}He, ^{35}Ar, ^{19}Ne). The accumulated data samples are large enough to provide a very high statistical precision in all three measurements and the preliminary analysis already confirmed the quality of the datasets. However, a thorough simulation of the whole experiment is mandatory to control the different systematic effects which constrain the precision that can be reached. Such a simulation is detailed hereinafter; more information on the theoretical context and on the LPCTrap device is available in reference [5].

2 Simulation package

The complete simulation package is being developed with the ambition to be both modular and general. The modularity aspect arises in the possibility for a user to plug-in a different given module easily, e.g. another β-decay generator than the one provided. The generality is ensured through the clear separation between the core simulation and the configuration inputs, i.e. it is possible to describe any setup with the proper configuration files. This package is divided into four stand-alone main modules : the β-decay generator (initial kinematics), CLOUDA (initial vertices inside the Paul trap), the recoil ion (RI) tracker and the β particle tracker. The proper simulation of the trapped ion cloud is one of the highest contributing systematic effect in LPCTrap [2]. A special effort is thus being made on this part. Before describing all of them, it is useful to introduce Bayeux [6], a package developed by the SuperNEMO collaboration. Bayeux's purpose is to provide to its users a high-level wrapper for GEANT4 [7], using specific separated submodules. This is done through a user friendly interface that allows the user to focus on the precise description of detection geometries, the study of the meaningful physical processes, and the complexity of the data analysis without the heavy code writing that usually comes along. Throughout the simulation package, the geometry is managed in different ways, depending upon the needs of a given module. For a realistic electric fields modeling, a fine description is required in order to include all possible contributions from even remote parts. The RI tracker and CLOUDA only need a description of the electrodes surrounding the ions of interest. Indeed, the low kinetic energy of both the trapped and the recoil ions (a few hundreds of eVs) implies that a single collision with any volume prevents them to reach the detector. They would thus be considered lost. The β tracker requires information about the precise geometry and material in which the surrounding volumes are made in order to precisely study the possible electron scattering, another important systematic effect at LPCTrap [2].

2.1 β decay generator

The aim of this module is to provide the kinematics of the initial decay for the RI and the β particle (and possibly the neutrino), including the Fermi correction, possible γ de-excitations and of course the $a_{\beta\nu}$ parameterization. The randomization loop is done with a Von Neumann algorithm to include the γ contribution. The level of experimental precision reached nowadays (below 1 %) is such that it may also require to take into account higher correction orders. In particular, radiative order-α corrections can reach the percent level in specific cases [8]. These effects are not included yet although a deep study of this aspect is required. Currently, the decay generator provides initial kinematics for the three nuclei of interest in LPCTrap but it can easily be extended to other elements.

2.2 CLOUDA

In order to correctly generate the position of the decay vertices within the confinement volume of the Paul trap, one needs to model the effects of the RF electric field, the collisions with the buffer gas and the impact of the space charge of up to a few hundreds of thousands of ions. The achievement of such a simulation with a reasonable amount of resource and a high precision is made possible using Graphical Processing Units (GPUs). CLOUDA is a homemade package running with CUDA and ROOT [9] and it is based on the N-body example provided by Nvidia [18]. Recently, CLOUDA began to yield its first results. These are preliminary and need to be tested against SIMBUCA which has been more widely adopted by the community [10]. It is worth mentionning that CLOUDA differs from Simbuca on four different points. Clouda implements a realistic buffer gas collision model, a realistic field with the usage of the harmonic synthesis, the n-body will be computed in the near future with a Barnes-Hut algorithm and all the calculations are performed on the GPUs whereas in Simbuca only the n-body effect is performed on the GPUs. From a technical point of view, CLOUDA has been working with four different GPUs (Tesla C2050, GTX670m, GTX760, Titan Black). Although the compatibility with ROOT 6 is still an issue, it has been running with different subversions of ROOT 5.34. The support for double precision real numbers is included and will work as long as the user's device is of at least compute capability 1.3[1] [11]. CLOUDA stepper is a basic Euler algorithm (thus a tiny time step was chosen), although it is foreseen that a more precise stepper will be implemented from the ones available : Runge-Kutta, Leapfrog, Verlet and so on.

2.2.1 Principles

Trapping field CLOUDA could include both electric and magnetic fields, although only electric fields support is available at the moment - the LPCTrap setup being based on a Paul trap. There are currently two ways to describe the confining field : to consider that it is ideal (faster) or to use an input of a specific set of harmonics and let the program compute a realistic synthesis out of this set (slower). In the ideal field case, the Poisson equation, the

---

[1]The *compute capability* is a hardware version clearly defined for each Nvidia's GPU card in its specifications.

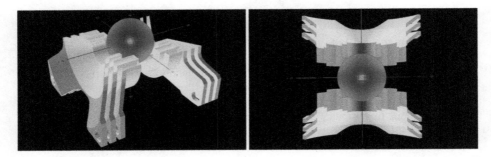

Fig. 1 The *central* part of the realistic geometry used to extract the harmonic coefficients in order to properly reproduce the confining field inside the Paul trap. The *red sphere* sitting in the center is the harmonic field description domain of definition

geometry constraints of the Paul trap electrodes and the solution of the equations of motion yield the following expression for the electric field [12, 13]:

$$\mathbf{E} = \frac{2 \times V_{pp} \times \sin(2\pi t f_{ion})}{R_0^2} (\mathbf{x} + \mathbf{y} - 2\mathbf{z}) \tag{1}$$

where V_{pp} is the peak-to-peak voltage applied on the electrodes with an ion-specific frequency f_{ion} for a Paul trap of radius R_0. This model being analytical, it is the fastest available. In the following, we will call $(2\pi t f_{ion})$ the radio-frequency phase (RFP). The harmonic model is separated into two distinct parts. First, one needs to extract the harmonic coefficients that will be used during the simulation. A detailed description of the whole setup has been used to compute the field in all space. The potential $V(\mathbf{r})$ (and/or $\mathbf{E}(\mathbf{r})$) itself is calculated using a homemade program based on indirect colocation Boundary Elements Method (BEM). It has been thoroughly checked against analytic formula and SIMION and allows a more detailed description of the experimental setup as it only requires a mesh of electrodes surface instead of a full space mesh. A volume of interest must be defined where the harmonic description will be applicable. In the LPCTrap case this is chosen as a sphere of radius $r_0 = 10$ *mm* around the center of the trap (see Fig. 1). $V(\mathbf{r})$ (and/or $\mathbf{E}(\mathbf{r})$) is computed for a large set of points on all the spherical surface. A fast Fourier transform is then applied to extract the list of harmonic coefficients contributing to the field. The second step consists in using these harmonic coefficients to synthesize a realistic potential at runtime, through the following equation:

$$\Phi(r, \theta, \varphi) = \sum_{\ell=0}^{\infty} \left(\frac{r}{r_0}\right)^{\ell} \sum_{m=0}^{\ell} P_{\ell}^m (\cos(\theta))(\mathcal{A}_{\ell m} \cos(m\varphi) - \mathcal{B}_{\ell m} \sin(m\varphi)) \tag{2}$$

where r, θ, φ are the ion spherical coordinates, r_0 is the chosen sphere radius and P_{ℓ}^m are the Legendre associated functions of first kind. In practice, the ℓ-sum is restricted to a given higher order ℓ_{max} which has been optimized to reach 10^{-5} precision on reconstructed field components. It is worth noting that this harmonic synthesis is valid inside the sphere of interest and becomes a poor approximation as ℓ_{max} decreases and as the field is computed near the sphere surface (especially near the electrodes). Taking LPCTrap as an example, a preliminary study for the two different field maps (ideal and harmonic) on all the RFP range (0 to 2π) yielded a maximum relative difference of about 10 %, located

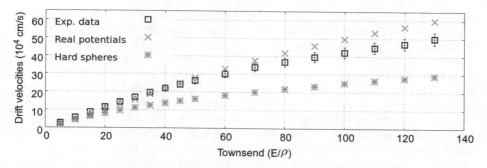

Fig. 2 Comparison between experimental drift velocities and simulated one using CLOUDA's collision models

on the edge of the chosen sphere. Such a difference will imply a different cloud shape, as shown below. During the experiment, the true radio-frequency applied on the electrodes does not follow exactly a sine curve. This observable is recorded on a event-by-event basis during the experiments performed with LPCTrap and needs to be included in the future field modeling.

Buffer gas collision A low pressure ($\sim 10^{-5}$ mbar) of buffer gas is required in order to cool down the confined ion cloud. The modeling of the collisions between the neutral buffer gas atoms and the charged ions has been an important part of this work. Two steps are taken to compute a collision: in the first one, the program determines whether a collision occurs or not (algorithm described in [14]) and if so, it computes the momentum transfer according to a specific model which provides the scattering angle (θ_{CM}) as a function of the available energy in the center-of-mass collisional frame.

For now, two models are fully implemented in CLOUDA : a faster classical hard spheres collision model (which works for any ion-atom couple of known Van der Walls radii) and a slower yet more accurate one, which employs diffusion data stemming from partial-wave collisional calculations and *ab initio* computations of binary interactions. The latter model will be referred to as 'real potential' in the following. Two other models are under development. The first is based on a Lennard-Jones potential and the idea of the second one is to define a scattering diffusion angle according to a specific probability distribution which is related to the mass difference between the buffer gas atom and the ion. The two working models were tested by computing drift velocities and comparing them with available experimental data [15, 16]. These data were gathered using Ar^+ ions produced and accelerated with a uniform electric field in a drift-tube containing a flow of neutral helium kept constant with a pump. Such a drift-tube is set in pulse mode to count arrival times of ion bunches thus yielding a drift velocity as described in [17]. In our simulation, we apply a uniform electric field on a given initial state and wait for thermal equilibrium to occur. The mean velocities of the simulated ion bunches are the drift velocities for the same E/ρ ratios as in the experiment (E being the electric field intensity and ρ the density of buffer gas). The results are shown in Fig. 2 where the drift velocities are plotted against different Townsend values. The real potential yields drift velocities values which are closer to the experimental ones compared to the hard spheres results. Although the difference might be considered rather large, it will be shown below that this seems to have no effect on the modeling of the trapped ion cloud.

N-body It is especially for the N-body interaction that the massive parallelization on GPUs is useful, as it is the most expensive effect to compute. The ions confined inside the Paul Trap are charged, meaning that they repel each other. While it is a simple Coulomb interaction, resolving the acceleration it induces is very expensive in terms of computing time. As said earlier, CLOUDA is based on the N-body example provided by Nvidia and as such, it kept its optimized brute force algorithm: the so-called Tile Calculation algorithm (detailed in [18]). Although it has optimized memory access and thread repartition, this algorithm still works in $O(N^2)$. In CLOUDA, the next steps will be to implement a Barnes-hut [19] ($O(N \log N)$) algorithm and a Fast Multipole Method (Specifically ExaFMM [20] which achieves $O(N)$ complexity). Both of them use an octree to recursively divide space, weighting the contribution of all charges according to their relative distance. The difference between these algorithms lies in the development of the contributions.

2.2.2 Cloud energy distributions

In order to study the effect of all possible interactions on an ion cloud in a specific configuration, it is required to reach a thermal equilibrium. An exponential fit of the form $a(1 - e^{-t \ln(2)/\tau}) + c$, with a, τ and c the fit parameters, is performed on the cloud mean kinetic energy as a function of time and the cloud is considered thermalized when reaching 10τ. The buffer gas collisions with the ions are responsible for the cloud thermalization and this destroys any information about the initial state. When the thermalization is reached, the cloud phase-space and mean kinetic energy still depend on the specific RFP being applied at a given time. All following simulations were done for ^{35,40}Ar using a ^{4}He buffer gas at room temperature. Simulations showed that changing the pressure only affects the time required to reach equilibrium. A pressure of 1 Pa was thus chosen to accelerate the simulations without changing the final results on the mean energy distribution at thermal equilibrium when compared to the real experimental conditions where the pressure is 10^{-5} mbar. The hard spheres and the ideal field, with the N-body switched off, are considered the default reference setting. The difference between this default setting and other possible cases is shown in Fig. 3 in terms of the thermalized cloud mean kinetic energy as a function of the RFP.

Field effect When simulating a cloud using a harmonic synthesis of the confining field, the cloud is hotter when thermalization is reached. This is especialy true at high extremas where the mean kinetic energy is 12 % higher. This is not negligible and indicates that the ideal field is not a good approximation. A realistic description is thus required and will increase the computing time. The time step was set such that the expected micromotion of the trapped ions is reasonably described.

Buffer gas effect With the ideal field description, changing the buffer gas model does not seem to alter much the final cloud mean kinetic energy. However the impact that the ~1 % difference has on the final value of $a_{\beta v}$ must be ascertained as it would allow to decrease computing time. Indeed, the interpolation inside the real potential table to describe the diffusion processes is time consuming compared to a simple randomization with the hard spheres model. The effect of the time step has been verified as well, with a convergence to 10 ns in the case of the hard spheres model and 1 ns for the real potential.

N-body effect The N-body space charge effect was studied within the scope of the ideal field description and the hard spheres collision model. Up to now, the study was limited

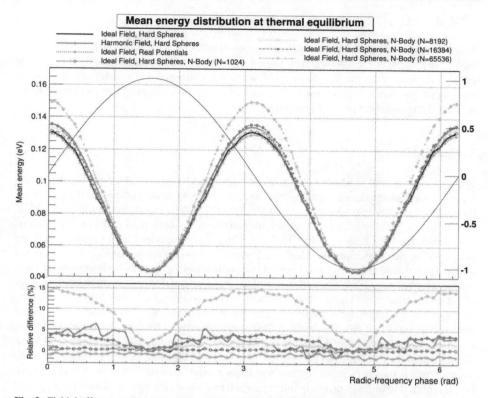

Fig. 3 Field, buffer gas and N-body space charge effect on the thermalized cloud mean kinetic energy as a function of the RFP. Smoothed lines are drawn to guide the eye. The *blue curve* is the applied radio-frequency phase. See text for details

for a number of ions equal to a power of 2 within the range [1024;65536]. Activating the N-body interaction does not seem to have an important impact for N=1024 ions whereas important differences for N=65536 emerge. The space charge effect has been tested with the same time step than for the two other interactions (buffer gas and trapping field), but further computer runs are necessary to reach convergence of the (optimal) time step.

2.2.3 RI tracking

The RI tracking is done using two different programs. The first one is included in the Bayeux package, where, as said above, GEANT4 provides the required Monte Carlo method. The second one is called SINS and is being developed at LPC Caen. The Bayeux RI part still requires a model of the electric fields in the relevant areas of the trap and the development of a plug-in to enable an arbitrary field description, while the geometry description is ready. SINS has the advantage of running on GPUs with powerful steppers and GPU hardware allocation tools, namely ODEINT (part of the BOOST library) and THRUST. SINS is not yet ready to track any ion, because the descriptions of the electric field and the geometry are still lacking. Both of them will be included through the usage of the texture memory available in CUDA which allows to use a high-performance on-the-fly interpolation.

2.2.4 β tracking

Another main systematic effect which affects the final TOF of the RI is the possible scattering of the β particle. GEANT4 and its embedded models will be responsible for this module through Bayeux. With the fine description of the geometry and the working tracker, everything is set to launch simulations of interest, although no TOF spectrum can be achieved without the completion of the RI tracking.

3 Conclusion

The full general and modular simulation package developed for LPCTrap has been presented. This package depends strongly on Bayeux (β-decay generator, part of the geometry, RI tracking, β particle tracking) and includes CLOUDA (ion cloud simulation with GPUs) and SINS (RI tracking with GPUs). Particular attention has been paid to the ion cloud modeling inside the Paul Trap with the CLOUDA stand-alone module. This simulation includes two working field descriptions (ideal and harmonic), two buffer gas models (hard spheres and real potentials) and the space charge effect computation with an N-body optimized algorithm. A preliminary study showed that the buffer gas model does not seem to have a strong effect on the final mean energy distribution at thermal equilibrium, while the harmonic synthesis of the confining field yields non-negligible difference when compared to the ideal field model (up to 12 %). While the N-body effect seems to be very small for 1024 ions, a non-negligible difference shows for 8192 ions ($\sim$ 1 %) and up to 65536 ions ($\sim$ 15 %). Knowing that the LPCTrap device trapped up to $\sim 2 \times 10^5$ ions in the experiments, the space charge effect is thus very important and will require to be taken into account. Although work is still required in order to properly integrate each module inside the simulation package and development efforts are still necessary, a reliable description (and concomitant optimization) of the whole LPCTrap experiments by means of the presented simulation package will be soon available.

Acknowledgments The authors acknowledge the computational facilities provided by the Mésocentre de Calcul Intensif Aquitain at the University of Bordeaux (http://www.mcia.univ-bordeaux.fr). They are also grateful to Samuel Salvador for his help and for making his hardware available.

Funding This work was supported in part by a PHC Tournesol (no. 31214UF).

Conflict of interest The authors declare that they have no conflict of interest.

References

1. Naviliat-Cuncic, O., Severijns, N.: Phys. Rev. Lett. **102**, 142302 (2009)
2. Fléchard, X. et al.: J. Phys. G: Nucl. Part. Phys. **38**, 055101 (2011)
3. Couratin, C. et al.: Phys. Rev. Lett. **108**, 243201 (2012)
4. Couratin, C. et al.: Phys. Rev. A **88**, 041403 (2013)
5. Liénard, E. et al.: Proceeding in this conference
6. Mauger, F.: The Bayeux library and the SuperNEMO software. GDR Neutrino Meeting, Orsay, June 18, 2014 (unpublished)
7. Agostinelli, S. et al.: Nucl. Instrum. Methods Phys. Res., Sect. A **506**, 250–303 (2003)
8. Glück, F.: Nucl. Phys. A **628**, 493502 (1998)

9. Brun, R., Rademakers, F.: ROOT - An Object Oriented Data Analysis Framework, Proceedings AIHENP'96 Workshop, Lausanne (1996); Nucl. Inst. Methods Phys. Res. A **389**, 81–86 (1997)
10. Van Gorp, S. et al.: Nucl. Instr. Methods A **638**, 192 (2011)
11. NVIDIA Corporation: CUDA C Programming Guide (2014)
12. Paul, W.: Rev. Mod. Phys. **62**(3) (1990)
13. Méry, A.: Ph.D. thesis, Université de Caen Basse-Normandie (2007)
14. Manura, D.: Scientific Instrument Services. Additional Notes on the SIMION HS1 Collision Model (2007)
15. Ellis, H.W. et al.: At. Data Nucl. Data Tables **17**, 177–210 (1976)
16. Lindinger, W., Albritton, D.L.: J. Chem. Phys. **62**, 3517 (1975)
17. Mc Farland, M., Albritton, D.L., Fehsenfeld, F.C., Ferguson, E.E., Schmeltekopf, A.L.: J. Chem. Phys. **59**, 6610 (1973)
18. Nyland, L., Harris, M., Prins, J.: GPU Gems, vol. 3, pp. 677–695. Addison-Wesley Professional (2007)
19. Barnes, J., Hut, P.: Nature **324**(4), 446–449 (1986)
20. Yokota, R., Barba, L.A.: GPU Computing Gems Emerald Edition, pp. 113–129. Elsevier, USA (2011)

⚉ Springer

Hyperfine Interact (2015) 235:97–106
DOI 10.1007/s10751-015-1184-2

The MR-TOF-MS isobar separator for the TITAN facility at TRIUMF

Christian Jesch[1] · Timo Dickel[1,2] · Wolfgang R. Plaß[1,2] · Devin Short[3] ·
Samuel Ayet San Andres[1,2] · Jens Dilling[4,5] · Hans Geissel[1,2] ·
Florian Greiner[1] · Johannes Lang[1] · Kyle G. Leach[3,4] ·
Wayne Lippert[1] · Christoph Scheidenberger[1,2] ·
Mikhail I. Yavor[6]

Published online: 8 May 2015
© Springer International Publishing Switzerland 2015

Abstract At TRIUMF's Ion Trap for Atomic and Nuclear Science (TITAN) a multiple-reflection time-of-flight mass spectrometer (MR-TOF-MS) will extend TITAN's capabilities and facilitate mass measurements and in-trap decay spectroscopy of exotic nuclei that so far have not been possible due to strong isobaric contaminations. This MR-TOF-MS will also enable mass measurements of very short-lived nuclides (half-life > 1 ms) that are produced in very low quantities (a few detected ions overall). In order to allow the installation of an MR-TOF-MS in the restricted space on the platform, on which the TITAN facility is located, novel mass spectrometric methods have been developed. Transport, cooling and distribution of the ions inside the device is done using a buffer gas-filled RFQ-based ion beam switchyard. Mass selection is achieved using a dynamic retrapping technique after time-of-flight analysis in an electrostatic isochronous reflector system. Only due to the combination of these novel methods the realization of an MR-TOF-MS based isobar separator at TITAN has become possible. The device has been built, commissioned off-line and is currently under installation at TITAN.

Proceedings of the 6th International Conference on Trapped Charged Particles and Fundamental Physics (TCP 2014), Takamatsu, Japan, 1-5 December 2014.

✉ Timo Dickel
t.dickel@gsi.de

[1] Justus-Liebig-University, Gießen, Germany

[2] GSI Helmholzzentrum für Schwerionenforschung, Darmstadt, Germany

[3] Simon Fraser University, Burnaby, Canada

[4] TRIUMF, Vancouver, Canada

[5] University of British Columbia, Vancouver, Canada

[6] Institute for Analytical Instrumentation, Russian Academy of Science, St. Petersburg, Russia

Keywords Isobar separation · MR-TOF-MS · ISOL · Low-energy RIB · Mass
measurements · Exotic nuclides

1 Introduction

The TRIUMF rare ion beam facility ISAC [1] currently serves 18 state of the art experiments. One of the experimental facilities at ISAC is TITAN [2, 3], a multi-ion trap system for precision experiments, such as mass measurements and in-trap decay spectroscopy. The main research fields are nuclear structure, nuclear astrophysics and fundamental symmetries and interactions. The latest upgrade to the TITAN facility is a multiple-reflection time-of-flight mass spectrometer (MR-TOF-MS). MR-TOF-MS have been installed recently at many rare ion beam facilities around the world [4–7]. It has been demonstrated that these systems can achieve outstanding performance, such as transmission efficiency up to 50 %, mass resolving power up to $m/\Delta m_{FWHM} = 600,000$, mass accuracies down to ~0.1 ppm, repetition rates up to 400 Hz, ion capacity in excess of $> 10^6$ ions per second and high sensitivity [8]. They can be used as highly accurate, fast and sensitive mass spectrometers or as isobar separators. At ISOL facilities like ISAC, mass measurements with an MR-TOF-MS enable access to shorter-lived nuclides and nuclides produced at lower rates than currently accessible with the standard Penning trap techniques. When the MR-TOF-MS is used as a mass separator, it facilitates measurements that are otherwise impossible due to isobaric contaminations. It has been shown that an MR-TOF-MS can be even used to provide isomerically clean beams [9]. In addition, the high mass resolution and capability to measure all isobars of one mass unit simultaneously enable a very efficient investigation and optimization of the target and ion source operation. Because of their versatility and compact and robust design they also find applications in other research fields such as in life sciences and in-situ analytical measurements [10].

2 Mass resolving power for mass separation at ISOL facilities

At ISOL facilities, the production yields of less exotic contaminants are often many orders of magnitude higher than the yields of the ions of interest, resulting in demanding requirements on the performance of the mass separator. This is especially true for ISAC, which has the highest power on target of all ISOL facilities world-wide.

The MR-TOF-MS offers an unparalleled combination of high ion capacity and mass resolving power [5] and is thereby the ideal solution for TITAN. In order to determine the necessary mass resolving power of the separator at an ISOL facility, the number of accessible nuclides were calculated for different mass resolving powers of the separator. The ISOLDE yield database was used for this investigation, because it has more entries than any other ISOL yield database. A nuclide was defined as accessible in this investigation if more than 50 % of the beam current after the separator corresponds to the nuclide of interest. The peak shape of the transmission spectrum of the separator was assumed to be Gaussian. The investigation was performed separately for the different ion sources, because each ion source provides a different composition of the beam. Only the four major ion sources surface ionization, plasma ionization, resonant laser ionization (RILIS) and forced electron beam induced arc discharge (FEBIAD) have been considered. To simplify matters the different energies, beam currents and targets have not been considered. The yields from surface ionization are added to the yields of the other three ion sources, because surface ionization

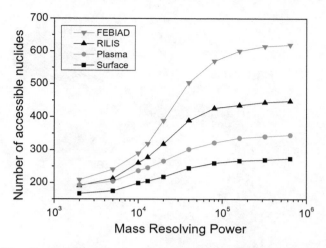

Fig. 1 The number of accessible nuclides for different ion sources in dependence of the mass resolving power (FWHM) of the separator at ISOL facilities, for details see text

always occurs and the production of the other ionization methods come in addition. After the yield database was divided into these four subgroups it was checked if more than one entry per isotope exists; if so, only the one with the highest yield was used.

In Fig. 1 one can see that a resolving power of 10,000 is necessary to significantly increase the number of accessible nuclides compared to the standard dipole magnet of ISAC, which has a mass resolving power of 2,000 [11]. For a mass resolving power of 20,000 about 70 % of the nuclides are accessible. To have more than 90 % accessible, a mass resolving power of about 50,000 is required.

3 Experimental setup

In Fig. 2, the layout of the TITAN facility is shown. The ISAC beam is captured, cooled and bunched in the radio-frequency quadrupole (RFQ) buncher [12]. There are numerous options for the experiment to proceed further: (i) transport of the bunched ions back downwards to other experiments, e.g. laser spectroscopy [13], (ii) direct transport to the measurement Penning trap (MPET) for mass measurements if no additional isobar separation is necessary, (iii) trapping in the electron-beam ion trap (EBIT) for in-trap decay spectroscopy or charge breeding and (iv) further transport of the charged-bred ions to the MPET for mass measurements. In the future the highly charged ions can also be cooled in the cooler Penning trap (CPET) before the mass measurement is done in the MPET. The MR-TOF-MS will enhance all operation modes (i-iv) by providing isobarically clean beams, and (v) it can be used as a mass spectrometer on its own to measure the most short-lived and rare nuclides.

The TITAN beamline is very compact and does not include any longer drift sections suitable for installation of the MR-TOF-MS. Thus the installation of the device "outside" the existing beamline as shown in Fig. 2 is the only option. This solution has the additional advantage that the system can be installed without major changes to the existing ion optics and beamline of TITAN, which is important since TITAN is a running facility.

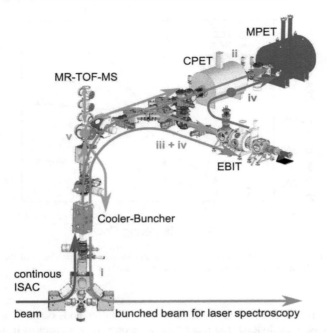

Fig. 2 Layout of the TITAN facility including upgrades (MR-TOF-MS and CPET) and the different operation modes, for details see text. The continuous beam from ISAC is shown in blue, singly charged ions bunched ions from the TITAN buncher are shown in red, singly charged ions processed by the MR-TOF-MS are shown in green and highly charged ions are shown in purple

Furthermore, the available space on the TITAN platform is very limited. A very compact device is necessary, because only a space of $0.8 \times 0.8 \times 1.5$ m^3 is available on top of the first 45° bender.

In conventional MR-TOF-MS isobar separators, the temporal separation is converted to spatial separation by a Bradbury-Nielson-Gate (BNG) behind the MR-TOF-MS [4]. The ions are then retrapped in an additional trap or they are transported electrostatically to the next experimental stage. This additional electrostatic beamline behind the MR-TOF-MS is not possible at TITAN due to the space restrictions discussed above. Two novel mass spectrometric methods had to be developed to allow for such a compact device: (i) Ion transport into and out of the device is performed using a buffer gas-filled RFQ-based ion beam switchyard [14]. The switchyard enables in- and ejection in all directions and merging of the exotic nuclides with ions from several off-line ion sources for calibration and optimization. The buffer gas-filled RFQ-based switchyard provides ion cooling and distribution without the need for additional differential pumping, thereby allowing a very efficient, compact, simple and reliable beam distribution. (ii) Mass selection is performed using dynamic mass-selective retrapping in the injection trap of the MR-TOF-MS after time-of-flight analysis in the isochronous reflector system. By closing the injection trap at a proper point in time, the spatial separation of ions is achieved, removing undesired ions and only storing the ions of choice. A detailed description and study of the mass-selective retrapping is given in [15, 16]. The system is based on the same analyzer as in [8, 17]. The combination of mass-selective retrapping and the RFQ-based switchyard allows to use the same trap and beamline for transport in and out of the system. An additional beamline can be omitted and

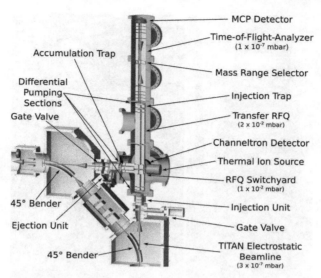

Fig. 3 Schematic Layout of the MR-TOF-MS at TITAN

a very compact system becomes possible. Thus, the combination of these novel methods allows the realization of an MR-TOF-MS based isobar separator at TITAN.

In Fig. 3 the schematic layout of the MR-TOF-MS at TITAN with its most important components can be seen. After the ions are ejected from the TITAN RFQ buncher at a potential of 30 kV, their kinetic energy is adjusted in a pulsed drift tube to about 1350 eV at ground potential, and they are transported to the electrostatic injection unit of the MR-TOF-MS. In the injection unit, the ions are steered, slowed-down and focused onto a set of vacuum separating apertures. In a gas-filled RFQ, the ions are cooled and can be stored. The RFQs of the MR-TOF-MS have a potential of about 1330 V. The ions travel through the RFQ switchyard, a transfer RFQ, and they are then cooled and bunched in the injection trap system. From the trap system, they are injected through two differential pumping stages into the time-of-flight analyzer [17], in which they travel with a kinetic energy of 1300 eV. A mass range selector (MRS) in the analyzer can be used to deflect ions that fall outside the desired mass range to ensure that the mass spectrum is unambiguous, i.e. all ions stored in the analyzer have undergone the same number of turns [8]. From the analyzer, the ions can be ejected either onto an MCP detector for measurement of their time-of-flight (e.g. for identification or mass measurement) or back into the injection trap system. After retrapping, the ions are cooled and sent through the transfer RFQ and the RFQ switchyard into the accumulation trap. Here the ions from several separation cycles are accumulated, thereby decoupling the operation frequency of the MR-TOF-MS (100 Hz) from that of the other TITAN components downstream of the MR-TOF-MS. The ions are ejected from the accumulation trap (potential of about 1280 V) into the EBIT or Penning traps. A channeltron detector and a thermal ion source are connected to the RFQ-based switchyard and provide diagnostic capabilities. At a later stage, a calibration ion source can be added as well. Ion transport from the accumulation trap back into the injection system and the re-injection into the analyzer for another consecutive separation cycle is possible allowing higher contaminant rejection if required. Ejection back into the TITAN RFQ buncher is possible as well in order to perform laser spectroscopy with isobarically separated ions. The MR-TOF-MS is connected to the

Fig. 4 The MR-TOF-MS in the laboratory in Gießen (*left*) and at TRIUMF (*right*)

existing TITAN vacuum system via gate valves, allowing independent operation of the TITAN facility as well as independent operation of the MR-TOF-MS. All ion optical elements, vacuum components and electronics are mounted in a single support frame to enable an easy transport, off-line tests and integration in the TITAN system.

The vacuum vessel of the MR-TOF-MS consists of DN160/250CF vacuum crosses. Three turbomolecular pumps (500 l/s) are used to evacuate the system. Encapsulated RFQs are employed in order to avoid additional differential pumping sections and pumps and thus to reduce cost and space requirements. Vacuum measurement of the pressures in the RFQs is performed using vacuum-compatible pressure gauges mounted directly on the RFQs in vacuum. The MR-TOF-MS uses highly stable power supplies (W-IE-NE-R Plain & Baus GmbH, Germany, Cologne and iseg Spezialelektronik GmbH, Germany, Radeberg). Custom-built circuits are used for HV stabilization, for generation of the RF voltages of the RFQs and traps and for HV pulsing. A digital storage oscilloscope is used for data acquisition.

For the commissioning before the installation on the TITAN platform the system has been equipped with an additional ion source (thermal Cs ion source, Heatwave Labs, USA, Watsonville) in front of the injection unit and an additional detector (MagneTOF, ETP Electron Multipliers, Australia, Clyde) behind the ejection unit. This allows the investigation of the beam transport through all components of the system.

4 Results

The system has been built, assembled and commissioned at the Justus-Liebig-University in Gießen, Germany. The MR-TOF-MS was shipped as an assembled system to Vancouver in September 2014 and re-commissioning starting in October 2014 at TRIUMF [18]. In Fig. 4

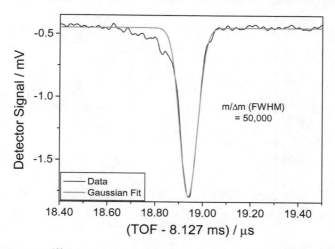

Fig. 5 Mass spectrum of ^{133}Cs$^+$ after 253 turns in the time-of-flight analyzer. The peak shape is almost Gaussian-like

photographs of the device in the laboratory in Gießen and at TRIUMF can be seen. Because the electronic circuits for floating the potential of the RFQs were not ready at the start of the commissioning, the ion kinetic energy had to be reduced to 650 eV for all measurements shown here.

As a first commissioning step, the ion transport from the external ion source to the MCP detector behind the time-of-flight analyzer and MagneTOF detector behind the accumulation trap was optimized. On the MCP detector, peak widths as short as 16 ns have been measured after 2 turns of the ions in the time-of-flight analyzer. The mass resolving power for large turn numbers has been investigated for ^{133}Cs$^+$ ions. A maximum resolving power of about 50,000 (FWHM) was measured in a mass spectrum after 253 turns in the time-of-flight analyzer. In the separator mode this mass resolving power would be sufficient to access more than 90 % of the nuclides produced (see Section 2). The peaks are in good agreement with Gaussian peak shapes (Fig. 5); they only show a weak tail on the left hand side. The tail is due to the particular tuning of the voltages of the MR-TOF-MS and the reduced kinetic energy, which leads to a larger beam diameter in the analyzer. In a next step, the mass-selective retrapping was tested. Resolving powers as large as 13,000 have been achieved (Fig. 6). The peak shape in separator mode shows less pronounced tails than the Gaussian peak shape. This is highly beneficial for the separation of ions with strongly different intensities. Because the MR-TOF-MS was operated at only half the design energy during the commissioning and since the time for the commissioning and optimization was limited, further performance improvements are expected in the future. However, already now the MR-TOF-MS provides a factor 6 higher resolving power than the dipole magnet currently used at ISAC and thereby almost 300 exotic nuclides for the different ion sources become now accessible.

5 Envisaged applications

The TITAN facility at TRIUMF offers superb possibilities for the research with exotic nuclides, particularly in the fields of nuclear structure, nuclear astrophysics and

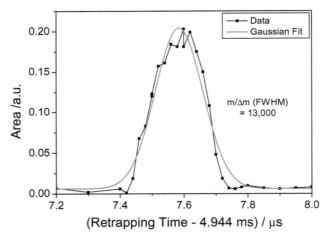

Fig. 6 Mass-selective retrapping of $^{133}Cs^+$ after 155.5 turns in the time-of-flight analyzer. The mass resolving power in separator mode demonstrated here is a factor 6 higher than currently available by the ISAC dipole magnet separator

fundamental symmetries and interactions. Many experiments have so far been hindered by strong isobaric contaminations. In the following, two examples will be given where the use of the MR-TOF-MS as isobar separator would be highly beneficial.

Exotic decay modes in the vicinity of proton-drip line for Z<30 [19] Of particular interest is β-delayed two-proton emission which was first predicted by Goldanskii [20] and was first experimentally observed for ^{22}Al [21]. Direct two-proton (2p) decay is another exotic decay mode. The direct 2p-decay process was first proposed theoretically by Goldansky [22]. Nuclides with a proton separation energy $S_p > 0$ and a two-proton separation energy $S_{2p} < 0$ are possible candidates for the two-proton radioactivity. The 2p-decay rate is extremely sensitive to S_{2p} and hence an accurate determination of this quantity is required [23]. Direct precise mass measurements will pin down the sign of S_{2p} and help to confirm experimentally the possibility of 2p-radioactivity. The measurements can be done by using the MR-TOF-MS as a separator and the Penning trap for the mass measurement or, in case of the most short-lived and weakly produced nuclides, the MR-TOF-MS will be used for the mass measurement.

In-trap decay spectroscopy For in-trap decay spectroscopy performed in the EBIT, isobaric contaminations result in increased background and complicated spectra [24, 25]. Thus these experiments will strongly benefit from isobarically clean beams provided by the MR-TOF-MS, which will result in an increase in sensitivity and accuracy.

6 Conclusions

The MR-TOF-MS at TITAN is based on novel mass spectrometric methods, the buffer gas-filled RFQ-based ion beam switchyard and the mass-selective retrapping. Only due to these, the installation on the TITAN platform becomes possible. The system has been commissioned at the Justus-Liebig-University in Gießen and installation at the TITAN facility is underway. The device can be used to optimize and monitor the production of the exotic nuclides, as an isobar separator and for mass measurements of the most short-lived nuclides. The system will facilitate many new opportunities by increasing the number of accessible nuclides at TITAN. A mass separation power of 50,000, which is sufficient to access more than 90 % of the nuclides produced at ISOL facilities, is within reach.

Acknowledgments   We would like to thanks T. Stora for discussions and providing the ISOLDE yield database. For excellent technical support we would like to thank A. Buers, M. Good, R. Weiß and C. Lotze. This work was supported by the Helmholtz Association of German Research Centers through the Nuclear Astrophysics Virtual Institute (VH-VI-417), by Justus-Liebig-Universität Gießen and GSI under the JLU-GSI strategic Helmholtz partnership agreement, and the German Federal Ministry for Education and Research (BMBF) under contract no. 05P12RGFN8.

References

1. Dombsky, M., Bishop, D., Bricault, P., Dale, D., Hurst, A., Jayamanna, K., Keitel, R., Olivo, M., Schmor, P., Stanford, G.: Rev. Sci. Instrum. **71**(3), 978 (2000)
2. Dilling, J., Bricault, P., Smith, M., Kluge, H.: Nucl. Instrum. Meth. B **204**, 492 (2003)
3. Dilling, J., Baartman, R., Bricault, P., Brodeur, M., Blomeley, L., Buchinger, F., Crawford, J., Lopez-Urrutia, J.R.C., Delheij, P., Froese, M., Gwinner, G.P., Ke, Z., Lee, J.K.P., Moore, R.B., Ryjkov, V., Sikler, G., Smith, M., Ullrich, J., Vaz, J.: Int. J. Mass Spectrom. **251**, 198 (2006)
4. Plaß, W.R., Dickel, T., Czok, U., Geissel, H., Petrick, M., Reinheimer, K., Scheidenberger, C., Yavor, M.I.: Nucl. Instrum. Meth. B **266**, 4560 (2008)
5. Plaß, W.R., Dickel, T., Scheidenberger, C.: Int. J. Mass Spectrom. **349**, 134 (2013)
6. Wolf, R.N., Errit, M., Marx, G., Schweikhard, L.: Hyperfine Interact. **199**, 115 (2011)
7. Schury, P., Okada, K., Shchepunov, S., Sonoda, T., Takamine, A., Wada, M., Wollnik, H., Yamazaki, Y.: Eur. Phys. J. A **42**, 343 (2009)
8. Dickel, T., Plaß, W.R., Becker, A., Czok, U., Geissel, H., Haettner, E., Jesch, C., Kinsel, W., Petrick, M., Scheidenberger, C., Yavor, M.I.: Nucl. Instrum. Meth. A **777**(21), 172 (2015)
9. Dickel, T., Plaß, W.R., Ayet San Andres, S., Ebert, J., Geissel, H., Haettner, E., Hornung, C., Miskun, I., Pietri, S., Purushothaman, S., Reiter, M.P., Rink, A.-K., Scheidenberger, C., Weick, H., Dendooven, P., Diwisch, M., Greiner, F., Heie, F., Knbel, R., Lippert, W., Moore, I.D., Pohjalainen, I., Prochazka, A., Ranjan, M., Takechi, M., Winfield, J.S., Xu, X.: Phys. Lett. B **744**, 137 (2015)
10. Dickel, T., Plaß, W.R., Lang, J., Ebert, J., Geissel, H., Haettner, E., Jesch, C., Lippert, W., Petrick, M., Scheidenberger, C., Yavor, M.I.: Nucl. Instrum. Meth. B **317**, 779 (2013)
11. Bricault, P., Ames, F., Dombsky, M., Kunz, P., Lassen, J.: Hyperfine Interact. **225**(1-3), 25 (2014)
12. Smith, M., Blomeley, L., Delheij, P., Dilling, J.: Hyperfine Interact. **173**(1-3), 71 (2006)
13. Mané, E., Voss, A., Behr, J., Billowes, J., Brunner, T., Buchinger, F., Crawford, J., Dilling, J., Ettenauer, S., Levy, C., Shelbaya, O., Pearson, M.: Phys. Rev. Lett. **107**(21), 212502 (2011)
14. Plaß, W.R. et al.: Physica Scripta **submitted** (2015)
15. Dickel, T. et al.: (to be submitted)
16. Lang, J.: PhD thesis, Justus-Liebig-University Gießen (in preparation)
17. Yavor, M., Plaß, W.R., Dickel, T., Geissel, H., Scheidenberger, C.: Int. J. Mass Spectrom. (2015). doi:http://10.1016/j.ijms.2015.01.002
18. Jesch, C.: PhD thesis, Justus-Liebig-University Gießen (in preparation)
19. Chaudhuri, A., Dilling, J.: TRIUMF EEC submission **S1333** (2011)
20. Goldanskii, V.: JETP Lett. **32**, 554 (1980)
21. Cable, M., Honkanen, J., Parry, R., Zhou, S., Zhou, Z., Cerny, J.: Phys. Rev. Lett. **50**, 404 (1983)
22. Goldanskii, V.: Nucl. Phys. **19**, 482 (1960)
23. Borrel, V., Jacmart, J., Pougheon, F., Anne, R., Detraz, C., Guillemaud-Mueller, D., Mueller, A., Bazin, D., del Moral, R., Dufour, J., Hubert, F., Pravikoff, M., Roeckl, E.: Nucl. Phys. A **531**, 353 (1991)
24. Lennarz, A., Grossheim, A., Leach, K., Alanssari, M., Brunner, T., Chaudhuri, A., Chowdhury, U., Crespo López-Urrutia, J.R., Gallant, A.T., Holl, M., Kwiatkowski, A.A., Lassen, J., Macdonald, T.D.,

Schultz, B., Seeraji, S., Simon, M.C., Andreoiu, C., Dilling, J., Frekers, D.: Phys. Rev. Lett. **113**, 082502 (2014)

25. Leach, K.G., Grossheim, A., Lennarz, A., Brunner, T., Crespo López-Urrutia, J.R., Gallant, A.T., Good, M., Klawitter, R., Kwiatkowski, A.A., Ma, T., Macdonald, T.D., Seeraji, S., Simon, M.C., Andreoiu, C., Dilling, J., Frekers, D.: Nucl. Instrum. Meth. A **780**, 91 (2015)

CPSIA information can be obtained
at www.ICGtesting.com
Printed in the USA
LVHW082318190619
621807LV00002B/16/P

9 783319 871158